THE SCIENCE OF LIGHT

BY PERCY PHILLIPS

D.Sc. (B'HAM), B.Sc. (LONDON), B.A. (CANTAB.)

LONDON: T. C. & E. C. JACK

67 LONG ACRE, W.C., AND EDINBURGH

NEW YORK: DODGE PUBLISHING CO.

CONTENTS

THE SCIENCE OF LIGHT

INTRODUCTION

This book is intended as a companion to that on *Radiation* previously published in this same series, and the first three chapters in that book should be read before commencing this.

Light plays such an important part in the life of man that it is only natural that speculation as to its nature and investigation of its laws should have begun very early. But although this is true, yet the ancients seem to have been exceedingly slow in the invention and construction of optical instruments, and to have been completely innocent of any reasonable optical theory. All the refinements both of construction and theory have happened well within the last three centuries.

Metallic mirrors are, of course, very ancient. Many of them have been recovered from ancient Egyptian tombs, and they are very distinctly mentioned in the books of Exodus and Job. Burning-glasses in the form of globes of water or of glass were also known at a very early time, Aristophanes mentioning them in a comedy of his which was performed in 424 B.C.

The most reasonable of the early theories was also one of the very earliest. Pythagoras, who died somewhere about 520 B.C., considered the sensation of sight to be caused by particles continuously shot out from luminous surfaces and entering the pupil of the eye. This is exactly the theory which Newton developed so brilliantly over 2000 years later, but the other theories were very fantastic and meaningless. Plato and his school maintained that the sensation of sight was caused by the union of three ocular beams. First was a stream of "divine fire" emanating from the eye itself; second, something emanating from the object seen; and third, the light of the sun. As far as one is able to form any concrete idea of his meaning, Plato looked upon these beams as something like tentacles feeling round until they came in contact with one another and then uniting, but it is doubtful whether he had a very clear idea of what he meant himself.

The Platonists knew two very fundamental laws of light, however. They knew that light travelled in straight lines when it travelled in a homogeneous medium, and they knew that when a ray of light was reflected at any surface, the angle between the incident ray and the surface was equal to the angle between the reflected ray and the surface.

About A.D. 150, Ptolemy, the Egyptian astronomer, investigated the refraction of light both by glass and water, but although he measured and tabulated the angles which the beam entering the glass or water makes with the surface, corresponding to certain angles which the incident beam makes with the surface, he was unable to see what law connected them. The next real advance was made by Alhazen the Arabian in the eleventh century. He described the anatomy of the eye, and showed how it produced an image of external things on the retina. He accounted for twilight and for binocular vision, and made real progress in the mathematical theory.

After this progress was very slow, until in 1608 Hans Lippershey, a spectacle-maker of Middelburg, constructed a telescope and published the discovery. A little later Galileo independently constructed another telescope, and a little later still, about 1615, Kepler showed how to calculate the magnifying power of the telescope from the focal lengths of the lenses employed. In 1621, Snell, who was then professor of mathematics at Leyden, discovered the laws of refraction which Ptolemy had failed to deduce from his experiments, but he died in 1626 without having published his results, and Descartes, having perused his papers, published it as his own discovery a few years later.

The latter part of the seventeenth century was a period of extraordinary activity and advance in the science of optics. In the fourteen years from 1665 to 1678 there were more important discoveries than had taken place in the previous fourteen centuries, and these discoveries mark the beginning of our present knowledge of the nature of light.

In 1665 appeared a treatise on light by Grimaldi, in which he gave an account of some interesting experiments on "diffraction," which is the name he gave to a small spreading out of light in every direction upon its admission into a darkened room through a small aperture. This spreading out shows that light bends round corners in the same way as sound does, but to a very much smaller extent.

In 1666 Sir Isaac Newton discovered the decomposition of white light into its component colours by means of a prism, and showed that no further colours were produced by a second refraction through a prism. He showed, moreover, that if the component colours be recombined they produce white light again. Newton's explanation of this was that white light is a mixture of the component colours, and that the sole function of the prism is to separate the components. Oddly enough this view, which we are taking as one of the most important beginnings of our modern knowledge of light, is one which we shall abandon in favour of the view that the prism actually manufactures the different colours out of the white light, and further, we have a pretty clear idea of how the manufacture is carried out.

Newton developed with remarkable ingenuity the idea that light consists of exceedingly minute particles shot out from a luminous body, and causing the sensation of sight when impinging on the retina. His handling of the theory was so masterly and his authority in the world of Science so great that he retarded the development of the wave theory for fully a hundred years;

yet, strangely enough, Newton himself made the discovery of what are known as Newton's rings, which, if handled properly, would have established the wave theory almost beyond the shadow of a doubt. Newton's rings are simply a special case of the well-known colours exhibited by thin films such as soap films, and will be considered in the chapter on interference.

In 1670 Bartholinus discovered the phenomena of double refraction and polarisation, but a hundred and forty-four years elapsed before Fresnel succeeded in explaining them.

In 1676 Olaus Römer, the Danish astronomer, by observing apparent irregularities in the time of rotation of Jupiter's satellites, demonstrated that light travels with a definite speed through space, and that the apparent irregularities are really due to the time taken by the light to travel from Jupiter to the earth. He estimated the speed by means of the measurements which then existed of the dimensions of the Solar system, and found it to be about 192,000 miles per second.

In 1678 Huygens, the Dutch physicist, formulated the first clear statement of the wave theory, which supposes light to consist of waves of some sort emanating from a luminous surface. He showed how reflection and refraction follow naturally from such a theory, but he was unable to show why light bends round corners so little. In consequence the theory made very little headway, and was not revived for over a hundred years.

After this wonderful period, 1665–1678, things advanced more quietly, and the next important discovery was that of the principle of interference by Young, early in the nineteenth century. He showed how a beam of light may be divided into two portions, which under certain conditions will produce darkness when both portions illuminate the same point. This follows quite naturally from any wave theory, but would be inexplicable by a corpuscular theory, and consequently the wave theory has supplanted the corpuscular theory from this time onwards. Young had in his mind the idea of longitudinal waves—that is, waves in which the vibrations of the individual particles are in the same line as the direction in which the wave is travelling. A simple illustration of a longitudinal wave is afforded by the transmission of a compression along a spiral spring. If one end of the spring is suddenly pushed forward, the end few coils of the spring are squeezed closer together. They recover by compressing the coils immediately in front of them, and so the compression is transmitted from end to end of the spring, and can be seen travelling unless the speed is too great. During the transmission of the wave, each coil has merely moved a little distance forward or backward along the length of the spring. The waves which constitute sound are compressional waves, like those in the spring. In 1814 Fresnel reintroduced a happy guess of Hooke's, made in 1672, and supposed the waves to be transverse—that is, the vibrations of the individual particles in the path of the wave are perpendicular to the direction in which the wave is travelling. Perhaps the most widely known examples of transverse waves are the waves on the surface of water. The individual particles of water move up and down nearly perpendicular to the surface, while the wave itself moves along the surface.

By assuming the transverse character of the waves, Fresnel was able to explain the polarisation of light, which had been discovered by Bartholinus so long before.

If light consists of waves, they must be waves in some sort of a medium, and since light travels across space in which there is no matter, we must suppose that the whole of space is filled with this medium. Much speculation as to the nature of this medium, which is called the ether, has been indulged in. Huygens himself conceived a kind of elastic solid, through which the vibrations of a luminous source are transmitted in much the same way as the vibrations of a marble embedded in a jelly are transmitted through the substance of the jelly. This idea served for some considerable time, and will of course still serve for those properties which light has in common with all other sorts of wave motion, but it breaks down when properties involving the actual character of the waves are considered.

The last great advance in our knowledge of the nature of light we owe to Maxwell, who in 1873 propounded his electromagnetic theory and showed that the ether which was required for the conveyance of light was the same as was required for the transmission of electric and magnetic actions, and that the known laws governing electric and magnetic actions would lead to electro-magnetic waves, which have all the characteristics of light.

Since that time there has been no radical change in the conception of the nature of light, but there has been steady advance along the lines laid down by Maxwell.

CHAPTER I

THE RECTILINEAR PROPAGATION OF LIGHT

The Propagation only approximately Rectilinear.—As the first stumbling-block to the wave theory of light was its inability to explain the approximate rectilinear propagation of light, let us consider this difficulty at the outset. The most ordinary observations on shadows show us that light travels in straight lines or rays, and this would seem at first sight to be strongly in favour of the corpuscular theory; but a closer examination shows that the rectilinear propagation is only approximate, and that light does bend round into the geometrical shadow of an obstacle in just the same way as sound does, only to a much smaller extent. Grimaldi had discovered this fact in 1665, as is mentioned in the Introduction, and any theory has therefore to explain an approximately, not an absolutely rectilinear propagation. The obvious superiority of the corpuscular theory therefore at once breaks down.

Huygens' Conception of Secondary Wavelets.—According to the wave theory, each point in a luminous surface is vibrating and sending out waves into the ether. We must therefore have some conception of the manner in which the disturbance is handed on from one portion of the medium to the next. Huygens' conception was as follows: "He regarded each vibrating point in the wave-front as the source of a secondary wavelet, and at any instant the surface which touches all the secondary wavelets is the new wave-front. Thus in Fig. 1 let O be a luminous point, and let PQ be a portion of the spherical wave-front. The neighbouring points *a*, *b*, *c*, *d*, *e*, . . . will all be vibrating in unison, and each of them may be considered as the centre of a new disturbance sending out its own spherical wavelet. At a certain instant these wavelets are represented by the dotted circles which have *a*, *b*, *c*, . . . as centres, and it is evident that the portion P′Q′ of another sphere with O as centre envelops the secondary wavelets and forms the new wave-front.

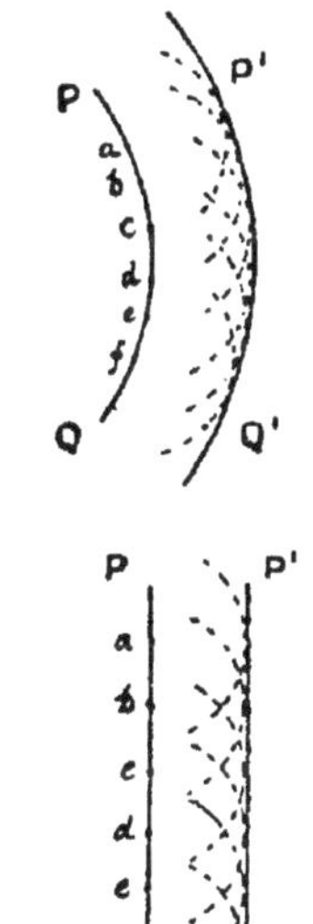

FIG. 1.

If the point O is a great distance away, PQ becomes a straight line and the wave-front becomes a plane wave-front. The secondary wavelets and the new wave-front are shown in the lower part of Fig. 1, and it is evident that the new wave-front is also plane. The effective part of the secondary wavelets would thus appear to be only that part which touches the enveloping surface. Huygens assumed this without any attempt at proof, thus assuming the whole question of the rectilinear propagation, and it was left to Fresnel to give a satisfactory explanation. Making use of the principle of interference discovered by Young a few years before, whereby it was shown that two sets of waves may destroy one another's effect, he showed that the absence of light in the shadow of an obstacle was due to interference between the secondary wavelets.

Real Existence of Secondary Wavelets.—The fact is that the secondary wavelets have a real existence and do spread out in all directions, as can readily be shown by means of ripples on the surface of water. Fig. 2 represents in plan a large shallow tank of water with a vertical screen standing up in it, the screen having a wide vertical slit in it at *a*. At the point O a disturbance sends out ripples in all directions, a convenient way of producing the disturbance being to move a pencil up and down in the water. With such an arrangement it is observed that the ripples which pass through *a* spread out from it as a new centre, just as indicated in the figure. The Huygens wavelets for sound waves have been shown very ingeniously by R. W. Wood. Sound waves consist of a series of compressions and rarefactions in the air or any other medium through which the sound is travelling.

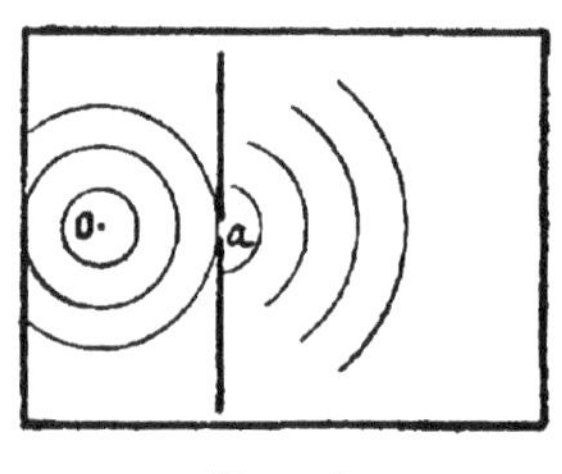

Fig. 2.

When the compressions and rarefactions follow one another perfectly, regularly a sustained musical note is produced. If the sound is a very short sharp one, *e.g.* a blow on wood with a hammer, the number of compressions and rarefactions is extremely small, and in some cases only a single compression may be produced. Wood produced a single compression by means of the crack of an electric spark. Any compressed or rarefied portion of air surrounded by undisturbed air will cast a fairly definite shadow of itself on a screen placed close behind it. For instance, if the shadow of a hot poker be received on a screen, the shadow of the hot air streaming upwards from the poker is seen quite distinctly. The hot air is less dense than the surrounding air, and consequently casts a shadow. If the poker be very cold instead of very hot, *e.g.* if it has been placed in ice or snow before being brought into a warm room, the shadow of the colder denser air can be seen streaming off downwards from the poker. Wood therefore arranged another spark so that it cast a shadow of the compression produced by the first spark on to a photographic plate. By adjusting the interval of time between the two sparks, the compression was allowed to travel different distances before being photographed and so show different stages in its development. In the path of the compression, two screens with slits in them were placed, and it was observed that the compression passing through each slit spread out from it just as if from a new disturbance. The following five diagrams are drawn from Wood's photographs. The

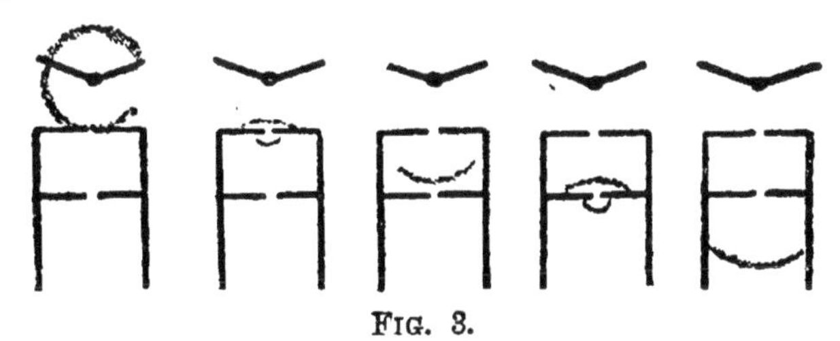

FIG. 3.

small black circle with the two lines coming obliquely from it at the top of each diagram represents the shadow of the two knobs, one behind the other, between which the spark which produces the compression passes.

The two horizontal lines with the small gaps in their centres represent the shadow of the two screens with slits in them placed in the path of the compression. The vertical lines are merely shadows of the framework holding the screens. The five photographs were taken with gradually increased intervals between the sparks, so that each diagram represents a slightly later stage than the one in front of it. In the first diagram the compression has reached the position of the shaded circle—that is, it has just reached the first slit. In the second a small portion of the compression has passed through the slit, and has spread out from it just as if the slit were a new source. The part of the wave which did not pass through the slit has been reflected back towards the sparking knobs. In the third the Huygens wavelet spreading from the first slit has just reached the second, and the reflected wave has travelled out of the field of view. In the fourth a Huygens wavelet has spread out from the second slit, and most of the first wavelet has been reflected back from the second screen. In the fifth the two wavelets have travelled on a little farther.

These two experiments show conclusively that in the case of both water ripples and sound waves any isolated element of the wave-front does act as a secondary disturbance, sending out the waves in all directions, while Grimaldi's experiments, which have already been mentioned, showed that light too spreads out from any small aperture in an exactly similar way.

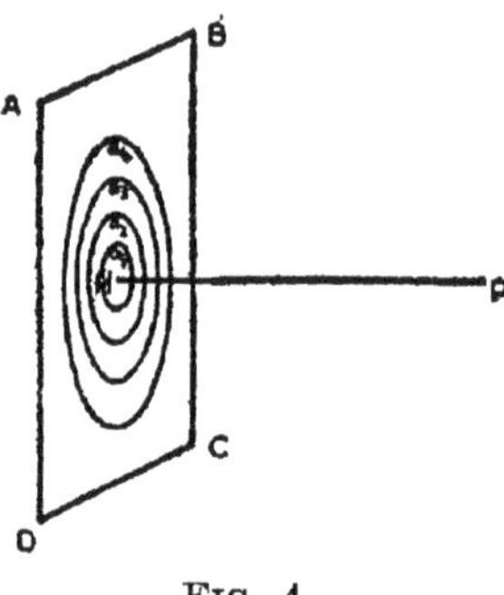

FIG. 4.

Half Period. Elements or Zones.—The illumination at any point in a wave-front is not, therefore, produced solely by that secondary wavelet which touches the wave-front at that point but is the sum of the effects of all the wavelets which reach that point from the wave-front. Thus if ABCD in Fig. 4 represents a portion of a plane wave-front which is moving from left to right, the illumination at P due to this wave-front is not due simply to the disturbance at N, but is the sum of the disturbances reaching it from all points in the plane ABCD. Fresnel estimated the sum of these disturbances in the following way: PN is drawn perpendicular to the plane, and with N as centre concentric circles a_1 a_2 a_3 . . . are described.

The radii are so chosen that Pa_1, is just half a wave-length of the light greater than PN, Pa_2 is half a wave-length greater than Pa_1, and so on. The ring between two consecutive circles is usually called a half-period element or zone. The name of Huygens' zones is often given to them for the apparently sufficient reason that Huygens had nothing whatever to do with them. If we consider the effect of two consecutive zones at the point P we see at once that the wavelets from a point in the outer ring will arrive at P just half a wave-length behind the wavelets from the corresponding point on the inner ring. They will therefore be just opposite in phase and will just neutralise each other if their amplitudes are equal. It is quite simple to show geometrically that the areas of the zones increase gradually as they are further from the centre, and so the amplitudes due to the outer zones would appear to be greater than those due to the inner ones. This is just counterbalanced, however, by the greater distance of the outer zones from P. Two consecutive zones would thus appear to neutralise each other completely, but there is one other circumstance to be allowed for, and that is the obliquity of the line joining P to the element in question, to the direction of propagation of the wave. It can be observed by inspecting Wood's photographs of the sound-waves, or by inspecting water ripples, that although the Huygens' wavelets spread out in all directions they are most intense in the direction of propagation of the original wave, and their intensity falls off gradually in directions more and more oblique to this. The final result is therefore that the effects of two consecutive half-period elements are very nearly equal and opposite, the outer having a slightly smaller effect than the inner one.

Summation of Effects of all the Zones.—Plotting a graph showing the amplitude at P, due to the different zones, we should get a curve something like Fig. 5, where the length RS represents the amplitude due to the fifth zone. Vertical lines are drawn as in the figure, and since the widths of the strips so formed are equal, the area of a strip can be taken to represent the effect due to a zone just as conveniently as the height of the strip. Since the effects of successive elements are opposite in phase we must attribute to successive strips opposite signs, and then the effect of the whole wave-front will be the algebraic sum of these strips. Folding the paper along the line BL we see that the point Q comes on to the point C and that therefore the algebraic sum of the first two strips is the triangle ABC. Repeating the folding along QM and the next vertical line we see that the algebraic sum of the next two strips is the triangle CDE, and repeating the process until the amplitude due to a zone becomes so small as to be negligible, we get a whole series of triangles extending from the top to the bottom of the first strip, each triangle representing the algebraic sum of the

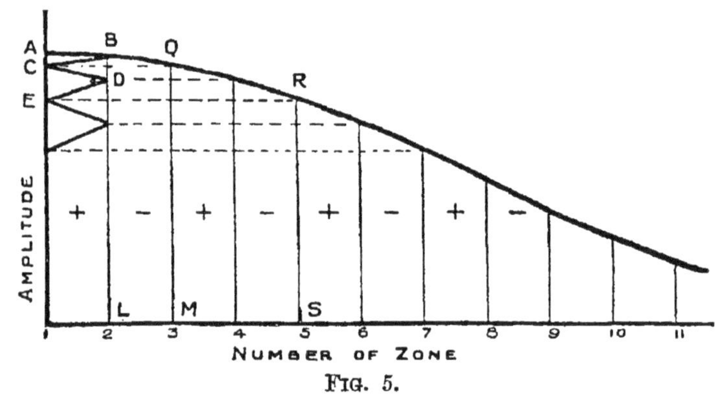

Fig. 5.

effects of two consecutive zones. The effect of the whole wave-front is evidently equal to the sum of the triangles, *i.e.* to one-half of the first strip. The whole wave-front therefore only produces half the effect of the first zone, and for most purposes we may think of the light as coming from the inner half of the first zone. It is also evident that when we have considered a good many of the zones in the centre the effect of the rest of them is negligibly small. Now the wave-length of light is so extremely small (between ·00004 and ·00007 cms.) that quite a small area round N (Fig. 4) contains a very large number of zones, and we therefore conclude that the effective portion of the wave is confined to this small area. The propagation of the light is therefore approximately rectilinear, in consequence of the extreme shortness of the wave-length, and is explained by combining Huygens' conception of secondary wavelets with the principle of interference.

Zone Plates.—A beautiful confirmation of Fresnel's theory is provided by what is known as a zone plate. This is a plane sheet of glass on which alternate half-period elements are blackened so that no light passes through them. The light which reaches the point P, in Fig. 4, from all the transparent zones is therefore all in the same phase, and brilliant illumination will result, just as if the zone plate were a convex lens. Estimating the amplitude due to these alternate zones, in the same way as is indicated in Fig. 5, we see that the resulting amplitude is the sum of the alternate strips, and as these are all of the same sign this sum will be very large. It is easy to show that the radii of successive zones are approximately proportional to the square roots of the natural numbers. To make zone plates therefore concentric circles are drawn on a large sheet of white paper with radii in this relation, and the alternate rings are blackened with ink. Several much reduced photographs of different sizes are then taken, and these provide zone plates with different sized zones. The size of the zones in Fig. 4 will evidently depend upon the distance of P from the plane, the greater the distance the larger being the zones. Each of the different zone plates will therefore correspond to a certain distance, PN, this distance being smaller the smaller the radii in the zone plate.

Illumination in the Shadow of a Circular Disc.—When Fresnel first expounded his theory to the French Academy, Poisson pointed out that if it were applied to the case of a circular disc illuminated by a point source of light on its axis it would lead to the result that the illumination on the axis in the shadow of the disc is the same as if the disc were absent, except for the slight reduction in intensity due to obliquity. Poisson thought that this was reducing the theory to an absurdity, and his reasoning was perfectly correct. The disc cuts off the light from the central zones, and it will be evident from an inspection of Fig. 5 that the effect of the remaining zones will be half that due to the first zone outside the disc. As a matter of fact the existence of a bright spot in the centre of the shadow of a circular disc had already been observed by Delisle, but it had been forgotten and was afterwards rediscovered by Arago and Fresnel. It can be observed using a small circular disc of metal such as a new threepenny-piece suspended by two fine threads attached to the coin with wax. As a source of light a pinhole in a thin sheet of metal is effective, some bright light being placed behind it. This source is placed three or four yards from the coin along its axis in a darkened room, and the shadow of the coin is explored with a low-power eye-piece or ordinary magnifying-glass three or four yards away from the coin along its axis on the opposite side from the source. A quite brightly illuminated region will be found all along the axis of the geometrical shadow. If we approach too near to the coin the illumination becomes feeble owing to the great obliquity of the secondary waves, but further away it is almost as bright as if the coin were absent. When the coin is viewed by an eye placed in the position of the bright spot the edge of the coin appears to be brightly illuminated, thus showing that the light in the central spot is coming from the edge.

CHAPTER II

THE REFLECTION AND REFRACTION OF LIGHT

Reflected and Scattered Light.—When light falls on any surface separating two substances part of it is always reflected back into the first substance. Thus when a narrow beam of light falls on to a polished mirror a narrow beam leaves the mirror along a definite path, and this path may be observed by projecting smoke into it. The particles of smoke scatter some of the light in all directions and so render the beam visible. The beam which follows a definite path is said to be regularly reflected, whereas another portion of the light which is scattered back in all directions is said to be irregularly reflected. This is rather a misnomer, as it is the surface which is irregular, not the reflection. The better polished a surface is the smaller is the amount of light which is scattered. The scattered light enables us to see most of the things around us which are not themselves luminous. Thus a polished piece of plate-glass is scarcely visible at all because so little light is scattered from its surface, but if the surface becomes dusty the light scattered from the dust makes it plainly visible. The scattered light is diffused in all directions and so can enter the eye in all positions and from all points of the surface.

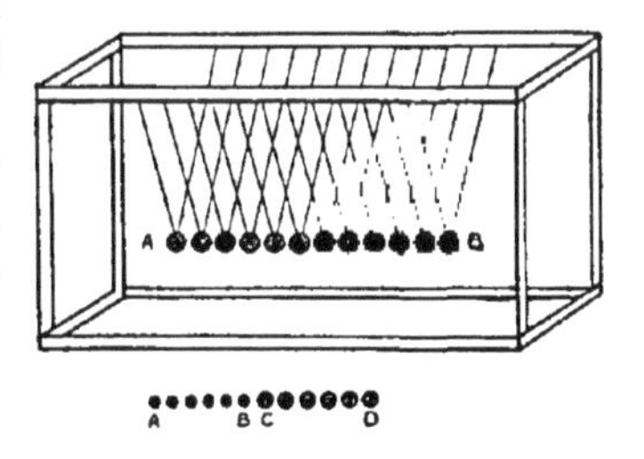

Fig. 6.

Model to illustrate Reflection.—The manner in which reflection is brought about may be illustrated by the model shown in Fig. 6, in which a number of exactly similar balls are suspended in a row from a framework so that they almost touch one another. The ball A is pulled out to the left and allowed to swing in and strike the ball next to it. This one moves forward and strikes the one next to it, and so on until the disturbance has travelled from A to B. Each ball will be practically stationary as soon as it has struck the one in front of it and so communicated its motion. There will thus be an absence of any reflected disturbance until the end B is reached.

Suppose now that the last half of the balls are replaced by heavier ones (C to D in the lower row). When the ball B strikes against C it will not be brought to a standstill as it would if C were equal to it, but will rebound from it and so start a disturbance which will be transmitted back to A. If the original disturbance be started at D instead of at A it will travel to C without any reflected disturbance, but when C strikes against

B it will not be brought to a standstill by the smaller ball but will follow on after the impact. It will then swing back and start a disturbance travelling back again to D. We may now imagine the two rows of balls to be two media whose surface of separation is at the point between B and C. When any disturbance originates in one of the media it is propagated through it until it reaches the surface of separation. At this point two disturbances are produced, one which passes on into the second medium, and the other which is reflected back into the first. From this point of view

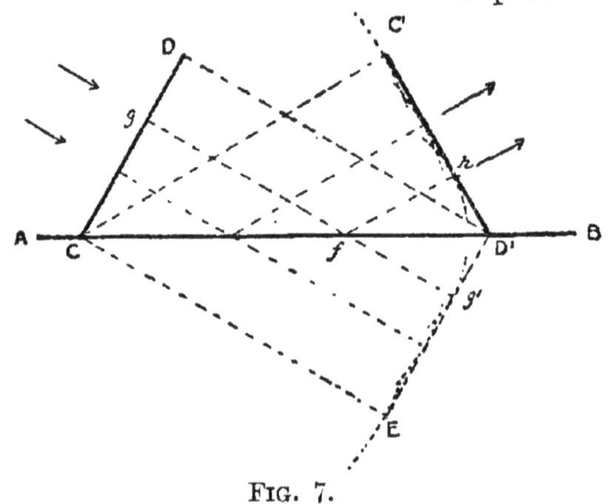

Fig. 7.

therefore each point in the surface of separation is considered as a new source of disturbance, and we may investigate the laws of reflection by constructing the Huygens' secondary wavelets from these points.

Reflection of a Plane Wave by a Plane Surface.—Let us take as our simplest case a plane wave incident on a plane surface and let Fig. 7 represent the section which is perpendicular to both the wave and the surface. AB represents the surface and CD represents the plane wave-front moving perpendicular to itself up towards the surface. If the surface had not been there it would have travelled in a certain time to the position ED′, but as each point in the wave reaches the surface it starts a secondary wavelet there. The secondary wavelet from C will therefore have travelled a distance CC′ which is equal to CE at the time when the original wave would have reached ED′. At the same time the secondary wavelet from f will have reached h where fh is equal to fg'. Treating every point in the surface in the same way and drawing the surface which touches the secondary wavelets we get the position of the reflected wave-front at this particular instant. If we consider the two triangles fhD′ and fg'D′, $fg' = fh$, fD′ is a common side, and the angles at h and g' are right angles, therefore the two triangles are equal and the angles hD′f and g'D′f are equal. But the angle g'D′f is equal to the angle DCD′, for the two lines DC and D′E are parallel, and therefore the angle hD′f is equal to the angle DCD′. This is true whatever secondary wavelet we consider, and therefore the straight line D′C will touch all the secondary wavelets and will constitute the reflected wave-front. The angles DCD′ and CDC′ are called the angles of incidence and reflection respectively, and we have therefore proved that the angle of incidence is equal to the angle of reflection. This is one of the laws of reflection. A little consideration of the construction employed in Fig. 7 also shows us that all the lines in it must be in the same plane and that therefore the incident ray, the normal to the surface at the point of incidence, and the reflected ray are all in the same plane. This is the other law of reflection. With these two laws as a starting-point it is a matter of pure geometry to investigate the reflection of any sort of wave-front at any sort of surface. This is done in books on Geometrical Optics, but it would perhaps be interesting to apply the wavelet method for constructing the reflected wave to one or two other simple cases.

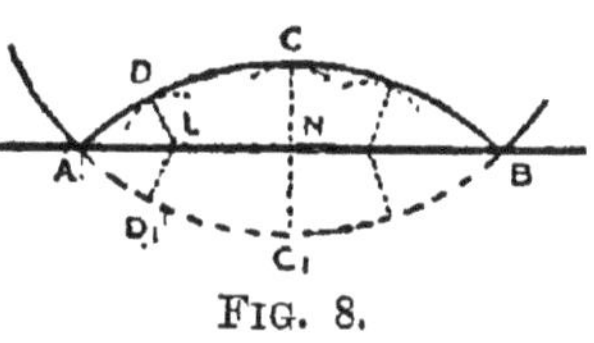

FIG. 8.

Reflection of a Spherical Wave by a Plane Mirror.— Let us take as a first case the reflection of a spherical wave-front at a plane mirror. Let O, Fig. 8, be the luminous point from which the spherical wave is diverging, and let the dotted part of the circle represent in section the position which the wave would have occupied at a certain instant if the mirror AB had not been present. After reaching the point N the wave would have travelled a distance NC′ and therefore the reflected wavelet will have travelled to C where NC=NC′. In the same way the wave after reaching L would have travelled a distance LD′ and therefore the secondary wavelet will have arrived at D where LD=LD′, and so on for all other points. It is evident therefore that if we were to fold the paper along the line AB the curve AD′C′B would lie along the curve ADCB. ADCB will therefore be a circle whose centre is where O would

FIG. 9.

fall when the paper is folded as suggested. That is, the centre would be just as far below the surface as O is above and would lie on the perpendicular drawn from O to the surface. This new centre is the point from which the waves appear to come after reflection and is therefore the image of the point O.

Reflection of Spherical Sound-waves at Plane Surface.—We have no means of directly showing the form of the wave-front of a luminous disturbance, and this is not surprising when we remember that light travels at the rate of about 186,000 miles per second, but R. W. Wood has photographed reflected sound-waves in just the same way as he photographed the Huygens' wavelets in sound. Fig. 9 is drawn from Wood's photographs. The small black circle with the two oblique lines from it in each diagram represents the shadow of the knobs between which the spark which causes the compression passes. The horizontal line is the shadow of the reflecting surface and the shadowy line represents the shadow of the compression itself. In the first position the unreflected and the reflected waves are in the field of view. In the second the unreflected wave has travelled out of the field of view and we only have the reflected wave. In the third the reflected wave has travelled a little farther. The form of the reflected wave is seen to be exactly that calculated from Huygens' construction.

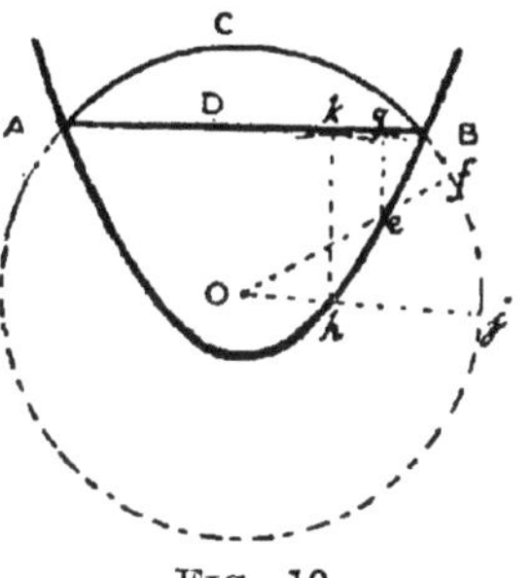

FIG. 10

Parabolic Mirror.—Another interesting case is the reflection by a parabolic mirror of a wave starting from its focus. We find that the reflection transforms the spherical wave into a plane one. Let ACB*fj*, in Fig. 10, be the position which a spherical wave starting from the focus O of the parabola would have attained at a certain instant. Draw two radii, O*f*, O*j* cutting the parabola at *e* and *h*, and from *e* and *h* draw perpendiculars to the straight line AB. It is a matter of simple geometry to show that *eg* is equal to *ef* and *hk* to *hj*, and therefore the secondary wavelets from *e* and *h* will have

FIG. 11.

reached *k* and *g* respectively. In the same way all the secondary wavelets will have reached the straight line ADB and therefore this represents the section of the reflected wave-front.

Reflection of Sound-wave at a Parabolic Mirror.—R. W. Wood has also shown the reflection of a sound compression at a parabolic mirror, and in Fig. 11 three successive positions of the wave are shown. In the first and second both the reflected and unreflected wave-fronts are shown, but in the third the unreflected one has travelled out of the field of view. Since the reflected wave-front is plane the rays, which are perpendicular to the wave-front, are parallel and emerge from the mirror in a narrow beam which does not spread out. This is the principle of the search-light, which consists of a brilliant light placed at the focus of a parabolic mirror. The reverse of this process is also true. Plane waves by reflection in a parabolic mirror are converted into converging spherical waves which shrink to a point at the focus of the parabola. This is made use of in

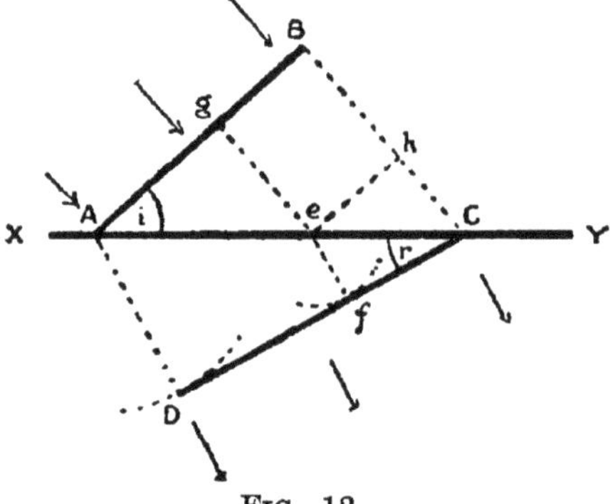

FIG. 12.

the reflecting astronomical telescope. Light waves coming from such distant sources as the astronomical bodies are practically plane and are therefore brought to a focus at the focus of a parabolic mirror. An image of the body is therefore produced there.

Refraction of Light. — When considering the model to illustrate reflection at the surface separating two media we saw that in general part of a wave which falls on the surface passes on into the second medium. This part is called the refracted wave. Let us first consider a plane wave falling on a plane surface of glass, and let us suppose that the speed of light in glass is less than it is in air. In Fig. 12, AB is the position at a certain moment of a plane wave travelling in the direction of the arrows up to the surface of the glass,

XY. While the wavelet from the point B is travelling in the air up to the surface at C the refracted wavelet from A is travelling in the glass. The distance BC which the one wavelet travels in air, divided by the distance AD which the other wavelet travels in glass, is equal to the speed of light in air, (v) divided by the speed of light in glass (v^1). In just the same way the wavelet from B travels from h to c, while the wavelet from g travels from e to f, and therefore hc divided by ef is also equal to $\frac{v}{v^1}$. Drawing the tangent plane CD to reach all the secondary wavelets, D, f, &c., we get the position of the refracted wave which will move on in the direction perpendicular to itself. We see at once that the refracted wave-front has turned so as to be at a smaller angle to the surface, *i.e.* its direction of motion is more perpendicular to the surface. Had the wave been travelling in the reverse direction, *i.e.* from a medium in which the speed is less into one in which the speed is greater, the reverse would have been true. It is usual to call the angle between the incident wave-front and the surface the angle of incidence (i), and the angle between the refracted wave-front and the surface the angle of refraction (r).

[1] Evidently $\sin i = \frac{BC}{AC}$ and $\sin r = \frac{AD}{AC}$.

Therefore $\frac{\sin i}{\sin r} = \frac{BC}{AD} = \frac{v}{v^1}$,

so that whatever the size of the angle i, $\frac{\sin i}{\sin r}$ is always the same. This is one of the laws of refraction derived experimentally by Snell in 1621, and the fraction $\frac{\sin i}{\sin r}$ *i.e.* $\frac{v}{v^1}$ is called the refractive index of the glass with respect to air. It is evident that when the wave passes from a medium in which its speed is greater into a

[1] See note at the end of the chapter.

medium in which its speed is less the refractive index is greater than unity, and vice versa. When the wave-front is parallel to the surface each point will reach the surface at the same time and therefore each point will start sending out a secondary wavelet into the glass at the same time. The refracted wave-front will therefore also be parallel to the surface and will go on undeviated.

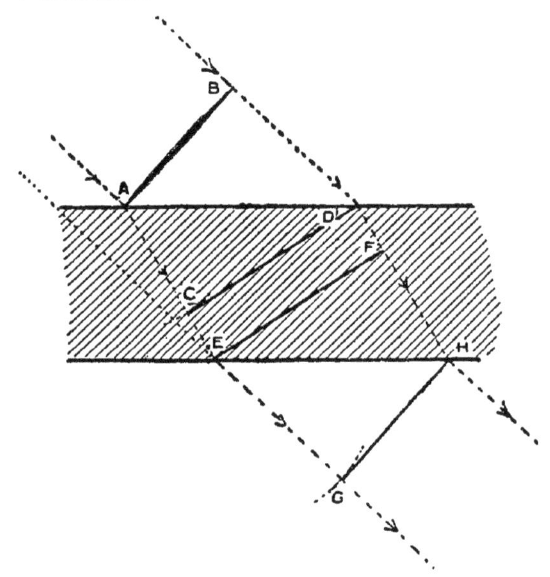

FIG. 13.

Refraction by Parallel Plate.—If a plane wave passes through a plate of the medium with parallel surfaces it is evident that at the second surface it will be deviated through the same angle into the original direction, the wave being simply displaced sideways by its passage through the plate. AB, CD, EF, and GH, in Fig. 13, represent successive positions of the wave-front passing through the plate, which is shaded. The amount of sideways displacement depends upon the obliquity of the wave to the surface as well as on the thickness of the plate. The displacement is evidently zero when the wave is parallel to the surface and is greatest when the obliquity is greatest, *i.e.* when the wave-front is almost perpendicular to the surface.

Refraction by a Prism.—If the second surface of the

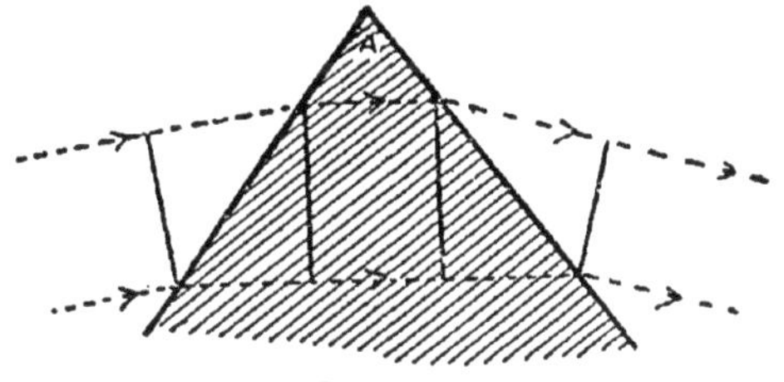

FIG. 14.

plate is not parallel to the first we have a prism. After passing through a prism the direction of the wave is altered as well as being shifted sideways, for the wave on reaching the second surface is not inclined to it at the same angle as it was to the first surface and therefore it will not be bent back by the same amount. Fig. 14 shows successive positions of a plane wave-front passing through a prism.

View of External Things from a Point under Water.—The fact that the deviation produced in a wave is greater

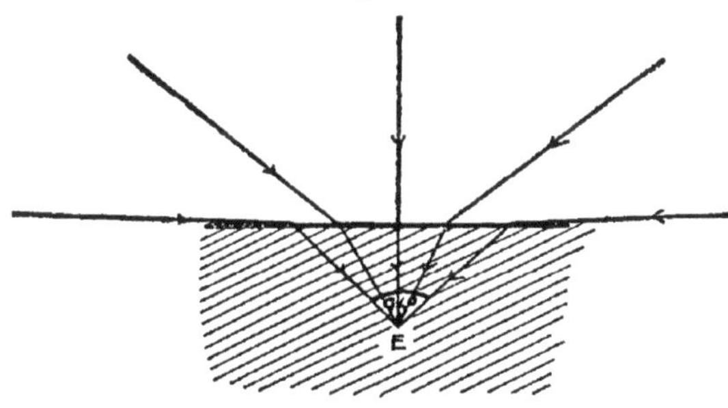

FIG. 15.

the greater the inclination of the wave to the surface has an interesting application in the views of external things which are obtained by an eye placed under water, *i.e.* the fish eye view of the world. In Fig. 15, E represents the point under the water from which external things are viewed, and the directions in which waves travel from different points in order to reach E are indicated by the lines with arrows. We see that all the rays reaching E are embraced by a cone whose angle is easily calculated from the refractive index of water to be 96°, and therefore the surface of the water appears to the eye at E to be an opaque roof in which there is a circular window immediately overhead, and into this circular window all external objects are compressed. Since the rays nearly perpendicular to the surface are very little deviated the objects immediately overhead are not distorted, but the nearer the objects are to the horizon the more will the rays be deviated and the more will the objects be distorted.

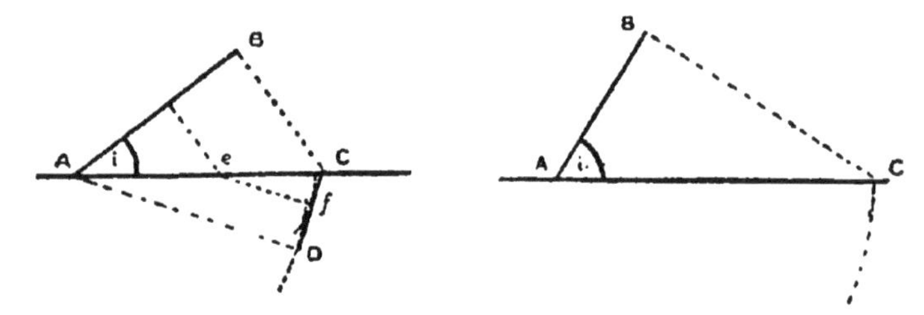

FIG. 16.

Total Reflection.—When a wave in one medium is incident on the surface of a denser one there is always a refracted wave possible, however large the angle of incidence may be, but if the wave is incident on the surface of a rarer medium a refracted wave is not always possible.

Suppose a plane wave-front AB, in Fig. 16, is incident on the surface AC of a rarer medium. Then CD represents the refracted wave-front where $\frac{BC}{AD}$ is equal to $\frac{v}{v^1}$ which in this case is less than unity. On increasing i we arrive at the point where $\frac{BC}{AC}$ is equal to $\frac{v}{v^1}$ and then the secondary wavelet from A will have travelled a distance AC when the wavelet from B has reached C. The plane from C touching the wavelet from A will therefore be perpendicular to AC and the refracted wave will graze the surface. If the wave-front AB is still more oblique to the surface the distance travelled by the wavelet from A while the wavelet from B travels to C would be greater than AC, and therefore it is impossible to draw a plane through C touching the wavelet from A. There is therefore no refracted wave-front, the whole of the wave being reflected at the surface. This phenomenon is known as total reflection.

Application of Laws of Refraction.—Having established the laws of refraction it becomes a matter of pure geometry to investigate the refraction of different forms of wave-fronts at different surfaces of different media, and these investigations are treated in books on Geometrical Optics. We will, however, show one or two simple results directly from the construction of the secondary wavelets.

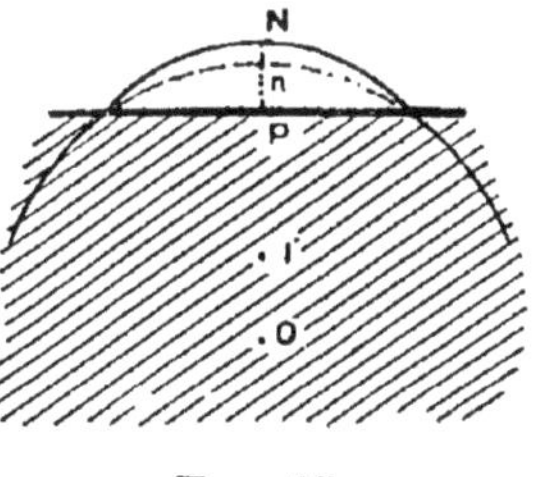

FIG. 17.

Spherical Wave at a Plane Surface.—Let us first consider the case of the refraction of a spherical wave-front at a plane surface. Let O, in Fig. 17, be the point from which the spherical wave starts in a dense medium, and suppose the position of the wave-front at a certain instant would have been the dotted circle if there had been no second medium. Then instead of the secondary wavelet from P reaching the point n it will have reached N where $\frac{\text{PN}}{\text{P}n}$ is equal to the speed of light in the second medium divided by the speed in the first. The wave-front will therefore become more convex and the centre from which it apparently comes will be at some point I which is higher up than O. The point O will therefore appear to be at I. This is why water appears shallower than it is, the real depth divided by the apparent depth being readily shown to be equal to the refractive index of the water if it is viewed nearly perpendicular to the surface.

Refraction of a Plane Wave by a Lens.—A convex lens is a portion of a medium which is bounded by curved surfaces usually spherical and which is thicker in the centre than at the edges. Let AOB, in Fig. 18, represent a plane wave coming up to such a lens. While the wavelet from O is travelling a distance OC through the material of the lens the wavelet from A will have travelled a distance AD, which is greater than OC because it has very little thickness of the denser medium to traverse. In the same way the wavelet from B will have reached E. As we consider points further and further removed from the centre O the wavelets will have less and less thickness of the lens to traverse and will therefore travel greater distances until the edges A and B are reached. The wave-front after traversing the lens will therefore be concave and will converge to some point F, which is called the principal focus of the lens. This type of lens thus causes waves to become more convergent.

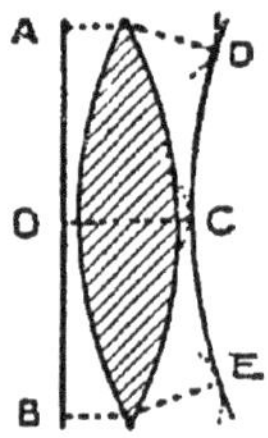

Fig. 18.

A concave lens is thinner in the centre than at the edges. Let AOB, in Fig. 19, represent a plane wave-front incident on a concave lens. The Huygens' wavelets from A and B will have to traverse a greater thickness of the lens than the wavelet from O, and therefore the wavelets in the centre will be in advance of those at the edges. The refracted wave-front will thus be convex and will diverge as if coming from some point F, which is called the principal focus of the lens. This type of lens therefore causes waves to become more divergent.

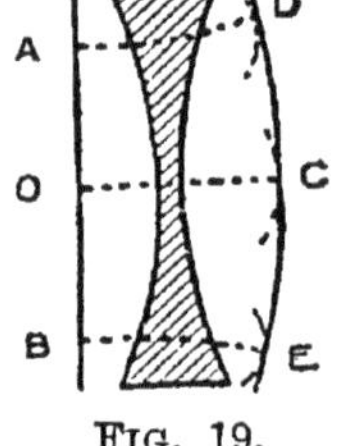

Fig. 19.

Refraction in Medium of gradually changing Density.—If a wave is passing through a medium whose density is gradually changing the direction of the wave will also gradually change. We have an example of this in the passage of light through the atmosphere. As the light travels from the upper rarer air into the denser air below its direction is gradually changed, and since the apparent position of an object is in the direction in which the light reaches the eye the heavenly bodies are all apparently displaced a little nearer to the zenith. Fig. 20 shows a wave-front AB travelling through the atmosphere. The point B being at a higher level than A the density of the air at B is less than at A, and the wave-front at B will travel faster than at A. This will happen at all points in the path of the wave, and therefore the end B will gain on the end A and the wave-front will become more horizontal.

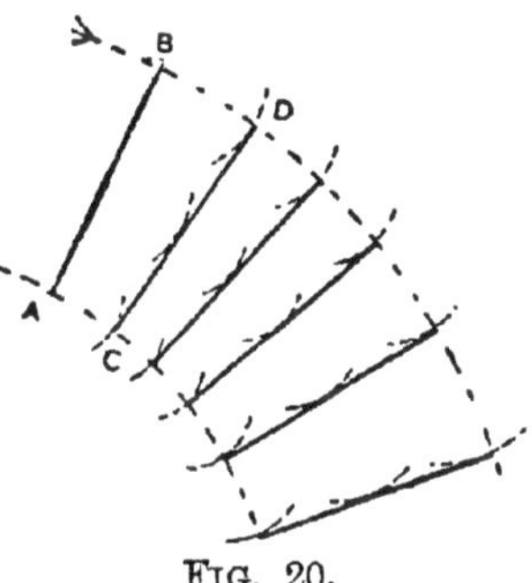

FIG. 20.

Mirage.—Another well-known example of refraction in a medium of varying density is the formation of the mirage. The air on the ground in contact with the hot sand of the desert becomes heated and so is less dense than the air above it. Waves of light coming obliquely downward into this layer will

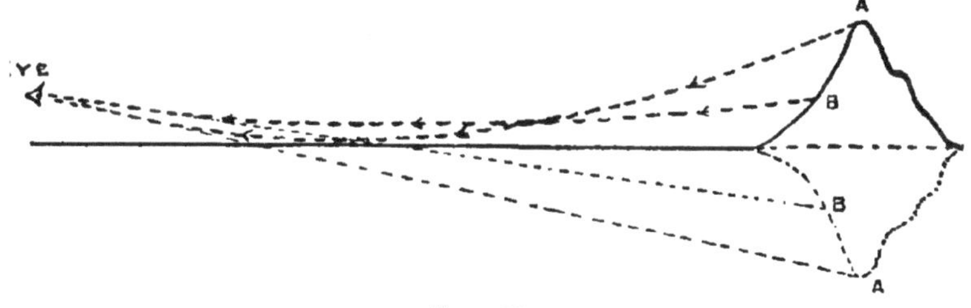

FIG. 21.

therefore be gradually turned upward from the ground, so that ultimately they are turned upward just as if they had been reflected in a sheet of water. The paths of two of the curved rays reaching the eye from two points A and B are shown in Fig. 21, and the rays are produced backwards in the directions in which they reach the eye, thus showing the apparent positions of the two points to be very like what they would have been if there had been reflection at a sheet of water.

Note on the Sine of an Angle.—In any right-angled triangle the side opposite to one of the angles divided by the hypotenuse of the triangle is called the sine of that angle, and if we denote that angle by A it is usually written sin A. On observing Fig. 22 it is evident that when A is small sin A is also small, for the side opposite to A is only a small fraction of the hypotenuse. As A approaches in magnitude to a right angle the side opposite to it becomes more and more nearly equal to the hypotenuse and so sin A becomes more nearly equal to unity.

FIG. 22.

If A is equal to a right angle the side opposite to it is the hypotenuse and therefore the sine of the angle is the hypotenuse divided by the hypotenuse, *i.e.* is unity.

CHAPTER III

DISPERSION

UP to the present point we have spoken as if a ray of light is all of it deviated by the same amount when it enters a second medium. Newton discovered, however, that when a ray of white light passed through a prism it was split up into coloured rays which were deviated by different amounts by the prism, and the white ray was thus spread out into a fan of coloured rays. A white screen placed in the path of these rays showed the well-known band of colours—violet, indigo, blue, green, yellow, orange, red. This band of colours is the spectrum of the white light. Inspection shows that the violet is the most deviated and the red least, and therefore we conclude that the violet travels the most slowly in the prism and the red the fastest.

This separation of the colours is called dispersion.

The Different Colours Travel with the same Speed through Space.—There is conclusive evidence to show that the different colours travel through interstellar space with exactly the same speed, for the variable star Algol shows remarkably rapid changes in brilliancy without any changes in colour. If the red travelled the smallest fraction faster than the violet it would arrive at the earth far in advance owing to the vastness of the distance from Algol to the earth. At the sudden increases in brilliancy of the star therefore it would first glow red, and at the sudden decreases in brilliancy it would become violet before fading.

Different Speeds of Different Colours in Refracting Substances.—We shall show later that the red light has the longest and the violet the shortest wave-length in the visible spectrum, and therefore in the normal prism the shorter waves travel more slowly than the longer ones. If we had a thickness of one mile of ordinary crown glass traversed by a beam of white light the red light on emerging would be about ten yards ahead of the blue light which entered at the same time. Flint glass would give a difference nearly twice as big. Any refracting medium has therefore a different refractive index for each colour in the spectrum, the refractive index for the violet being the greatest and for the red the least. There seems to be no very simple relation between the refractive index and the wavelength in any material, some media dispersing the colours much more widely than others for the same average deviation. This difference in dispersion by prisms of different kinds makes it possible to choose two prisms, such that the dispersion of the one is equal to the dispersion of the other, but the average deviation by one is greater than that by the other.

Achromatic Prisms.—If we place these two prisms close together with their angles turned in opposite directions a ray passing through them will be deviated without being spread out into a spectrum. A combination of this sort is called an achromatic prism and its mode of action can readily be followed by taking a numerical example. We have mentioned above that flint glass disperses the colours nearly twice as much as crown glass. If we fix on definite colours in the spectrum we can find the refractive indices of the two kinds of glass for each of the different colours. Let us fix on what are known as the C, D, and F lines in the solar spectrum. The C line is a deep red colour, the D line a yellow, and the F line a blue-green.

The refractive indices are:

	C.	D.	F.
Flint glass . .	1·630	1·635	1·648
Crown glass . .	1·527	1·530	1·536

If we use a prism of small angle, A, it is easily shown that the angle through which a ray is deviated on passing through the prism is equal to $(\mu-1)A$ where μ is the refractive index of the ray.

For such a prism of flint glass the C line would be deviated through an angle ·630A, and the F line through ·648A, so that the angle between the two rays would be ·018A.

Now suppose we have a crown glass prism whose angle is twice as great ($=2A$). The C line would be deviated through an angle $\cdot527\times2A=1\cdot054A$, and the F line would be deviated through $\cdot536\times2A=1\cdot072A$, so that the angle between the two rays would be again ·018A. Placing these two prisms with their angles in opposite directions, as in Fig. 23, the crown glass prism will deviate the C ray through 1·054A, and the flint glass prism would deviate it back again through ·630A, so that there would be a nett deviation of ·424A. The F ray would be deviated through 1·072A by the crown glass and through ·648A back again by the flint glass, giving a nett deviation of ·424A. The C and F rays will therefore emerge parallel, both having suffered a deviation of ·424A.

The deviation of the D ray would be $1{\cdot}060A - {\cdot}635A$, *i.e.* ·425A. This is very nearly equal to the deviation of the C and F rays, but is not exactly, and so although there is no dispersion between the C and F rays there is a very small amount of dispersion between the other rays. For most practical purposes this dispersion is

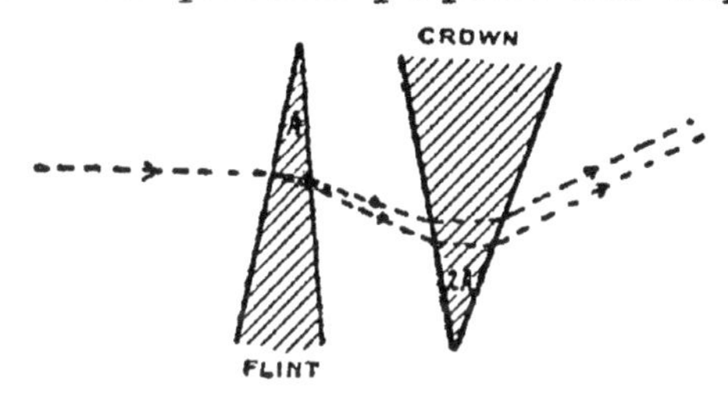

FIG. 23.

small enough to be negligible, but it is possible to arrange three prisms of different dispersive power so that three colours have exactly the same deviation. Even then the deviation of the remaining colours would not be exactly the same, but the dispersion would be very much smaller than with the two prisms only.

Dispersion in Lenses.—Achromatic prisms are of no practical use, but the principle is of enormous importance in the making of achromatic lenses. Owing to the difference in the refractive indices of the different

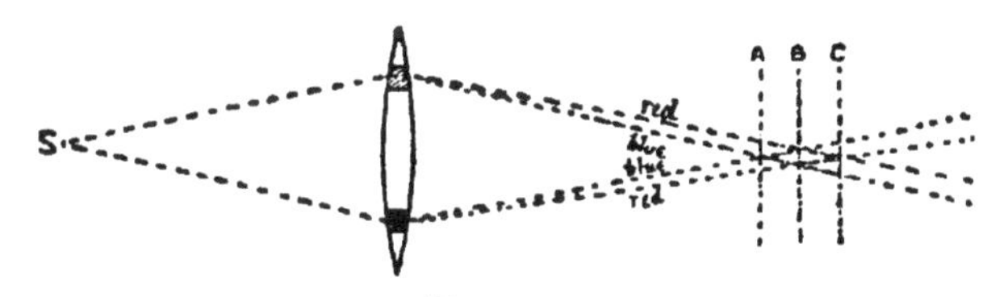

FIG. 24.

colours the images formed by single lenses are always imperfect and coloured. This can very readily be shown. Let S, in Fig. 24, be a point from which rays of light are diverging to a lens, and let us fix our attention on two of the rays. The two regions of the lens through which the rays pass are shaded in the figure in order to bring out the fact that each portion of the lens may be considered as a small prism. Since the blue end of the spectrum is the most deviated the blue rays will cross one another, *i.e.* will be brought to a focus at a point nearer to the lens than the red rays. Thus the blue rays will form an image of the point S at A, and the red rays at C. At A the image will have a red edge, and at C a blue one, while at B we shall get the best image, for there the rays pass through the smallest space. The image is far from perfect, however, for the image of S, instead of being a point, is a small patch of light. This defect in lenses is known as chromatic aberration.

Achromatic Lenses.—In order to correct for it we must use a combination of two lenses so that each portion of the combination can be considered as two opposed prisms just like the crown and flint glass prisms. Such a combination is shown in Fig. 25. On the left is a crown glass lens whose surfaces are both convex and of equal radii. On the right is a flint glass lens, one of whose surfaces is flat and the other concave, and of the same radius as the surfaces of the other lens. At any point in the combination the slope of the one surface of the flint glass lens will be the same as that of the crown glass; but since the other surface of the flint glass is flat, the angle between the two surfaces of the flint glass will only be half that between the two surfaces of the crown glass. At any point in the combination, *e.g.* the shaded region, we therefore have two opposed prisms of crown and flint glass, the angle between the surfaces of the crown glass prism being twice the angle between the surfaces of the flint glass prism. We have seen that under these circumstances all the colours are deviated by practically the same angle and therefore

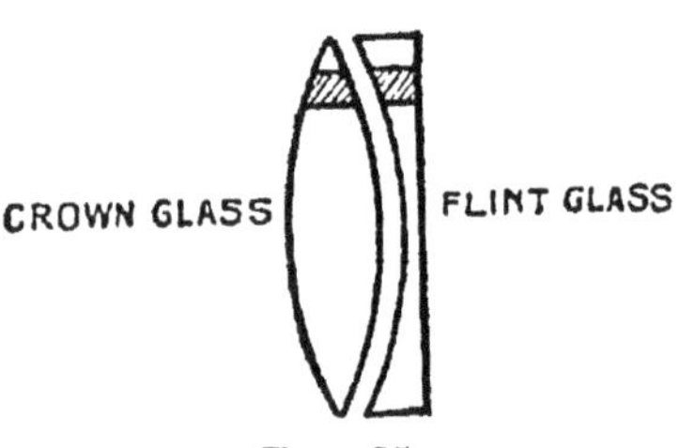

FIG. 25.

they will all be brought to a focus at the same point. Such a combination is called an achromatic combination.

Direct Vision Spectroscopes.—Just as it is possible to construct a combination of prisms which gives a deviation of the rays without dispersion, so it is possible to construct a combination which produces no deviation of the middle of the spectrum, but still produces dispersion. This is the principle of the direct vision spectroscope, which is useful because of its compactness.

A numerical example with opposed crown and flint glass prisms will again make the action clear. The refractive index of the ray midway between the C and F lines for flint glass is 1·639, and for crown glass 1·5315. If, therefore, the angle of the flint glass prism is $\frac{\cdot 5315}{\cdot 639}$ of the angle of the crown glass, the two prisms would deviate the middle ray by the same amount and in opposite directions.

Call the angle of the crown glass prism A, then the angle of the flint glass will be $\frac{\cdot 5315}{\cdot 639}A = \cdot 832A$.

The C ray will be deviated by ·527A by the crown glass, and by $\cdot 630 \times \cdot 832A = \cdot 524A$ by the flint glass. The nett deviation is therefore $\cdot 527A - \cdot 524A = \cdot 003A$.

The F ray will be deviated by ·536A by the crown glass, and by $\cdot 648 \times \cdot 832A = \cdot 539A$ by the flint glass. The nett deviation is therefore $\cdot 536A - \cdot 539A = -\cdot 003A$. That is, the C ray will be deviated in one direction and the F ray in the other, and the angle between them will be ·006A.

Normal Dispersion.—The dispersion in a prism is said to be normal if the refractive index, and therefore the deviation, is greater the smaller the wave-length of the light used. Even with substances whose dispersion is normal, however, the rate of change of refractive index with the wave-length is very different for different substances and in different parts of the spectrum. Thus some prisms will have the colours very closely crowded together at one end of the spectrum and widely dispersed at the other, while other prisms may crowd the opposite end of the spectrum.

Anomalous Dispersion.—Some prisms, however, will deflect some of the shorter waves by less than some of the longer ones and so the spectrum becomes mixed. This phenomenon is called anomalous dispersion, but we shall see that it is not really anomalous at all. On the contrary, the normal dispersion is merely a special case of the anomalous.

Crossed Prisms.—A simple way of studying the phenomenon is by the method of crossed prisms. In this method a spectrum of a bright point source of light is produced by a normal prism, with its edge placed

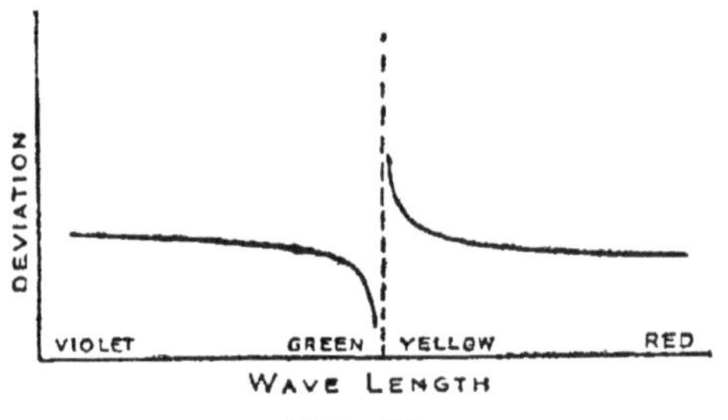

FIG. 26.

vertically. This would produce a narrow horizontal band of colour showing the colours in the normal order. In the path of the rays emerging from this prism a second prism is placed, but this time with its edge horizontal. This deviates the horizontal spectrum in the vertical direction and as the different colours are deviated by different amounts the band of colours will be either oblique or curved. If the second prism produces anomalous dispersion the band of colours will have a break in it at a certain wave-length and will appear something like Fig. 26, which is the curve for a cyanine prism. Just below a particular wave-length the deviation by the second prism is much smaller than for just above that wave-length, so that on the side of smaller wave-lengths the band of colours bends down as it approaches the particular wave-length, while on the side of greater wave-lengths it bends upwards. The particular wave-length towards which the bend occurs is always very strongly absorbed by the prism.

A very good prism for studying the phenomenon is made by squeezing a little fused cyanine between two plates of glass which are inclined at a very small angle. Although cyanine absorbs light very strongly, enough light will be transmitted through the thin edge of the prism to exhibit the phenomenon.

Even without using the crossed prisms the anomalous dispersion is noticeable, for one has only to view a slit illuminated with white light or even a flat gas flame turned edgewise to notice that the colours are in the unusual order--green, blue, red, orange. A glance at Fig. 26 shows us that this is what we should expect, as the green is deviated least and the orange and yellow most.

Normal Dispersion a Special Case of Anomalous Dispersion.—When the transparency of substances for waves outside the range of the visible spectrum is investigated it is found that substances which are quite transparent for all the visible spectrum, and which have perfectly normal dispersion for the visible rays, have strongly marked absorption bands in other parts of the spectrum, and the dispersion near to the absorption band is similar to that in Fig. 26. Some substances have several absorption bands and at each of them the dispersion has the same characteristics. We therefore conclude that all substances have absorption bands, *i.e.* they absorb certain waves very strongly, and that the general character of the dispersion is the same for all. Waves a little shorter than the absorbed ones are deviated abnormally little, and waves a little longer abnormally much. At a considerable distance from an absorption band the deviation gradually increases as we pass from longer to shorter waves, *i.e.* from red to blue.

An inspection of Fig. 26 also leads us to suppose that if the dispersion curve in the visible spectrum is concave upwards, we may expect an absorption band in the shorter wave-lengths, *i.e.* in the ultra-violet, whereas if the curve is concave downwards we may expect an absorption band in the longer waves, *i.e.* in the infra-red. These expectations have been realised in all the cases which have been investigated.

The Flash Spectrum.—A very beautiful example of anomalous dispersion is provided by what is known as the flash spectrum of the sun. Just as an eclipse of the sun becomes total, instead of the usual spectrum of the sun consisting of a continuous band of colour crossed by narrow dark lines, the continuous band fades away owing to the moon interposing, and the dark lines flash out bright on a dark background. To explain this we must remember that the sun consists of a white hot molten interior surrounded by an atmosphere of hot vapours most of which are metallic. In the chapter on the meaning of the spectrum in the companion book on *Radiation*, it is shown that the existence of the dark lines is explained by the absorption from the white light coming from the interior of the sun of those waves which are characteristic of the metallic vapours in the atmosphere. In the earlier explanation of the flash spectrum it was supposed that as soon as the very brilliant light from the interior of the sun was stopped by the moon the light from the atmosphere itself just passed the edge of the moon and so gave the characteristic bright line spectrum of incandescent gases. This would hardly suggest anything so brilliant as the flash spectrum, however, for the bright lines flash out very brightly indeed. W. H. Julius of Amsterdam a few years ago suggested that the light in the flash spectrum came from the interior of the sun and was abnormally refracted by the atmosphere of the sun. We have seen in the chapter on refraction (p. 33) that a ray passing through a medium of varying density such as our own atmosphere or the atmosphere of the sun will pursue a curved path. Since the refractive index of a gas is usually very little different from unity, for most waves the actual bending is very small indeed, but if the waves have a frequency very little greater than the characteristic waves absorbed or emitted by the gas the refractive index is abnormally high, and the bending may be considerable. Thus the direct rays from a point A, Fig. 27, on the liquid interior of the sun may be just intercepted by the moon from the point B on the earth. But a wave extremely near to one of the absorption bands of the sun's atmosphere may pursue a curved path like that indicated and reach the point B. Thus the wave reaching B from the interior of the sun will consist of waves extremely near to the waves which are characteristic of the gases composing the sun's atmosphere.

The production of the flash spectrum has been very prettily imitated by R. W. Wood, who used a sodium

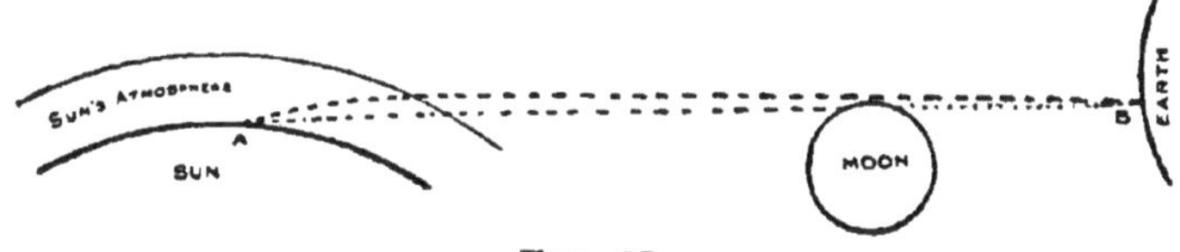

FIG. 27.

flame to represent the sun's atmosphere and a brightly illuminated plaster plate to represent the sun's interior. A brilliant flash spectrum consisting of the two sodium lines was observed, which disappeared almost entirely as soon as the light was cut off from the plaster plate, thus showing conclusively that the light came from the plaster plate and not from the sodium flame itself.

Theory of Dispersion.—The theory of dispersion is much too complicated to be treated in this book, but the foundation on which the theory is worked out may be suggested. In the little book on *Radiation* previously referred to an attempt has been made to show that light and heat consist of electromagnetic waves. It was also mentioned that the atoms of all substances are believed to contain numbers of small electric charges (the electrons) which are held in the atom by the electric attraction of other opposite charges within the atom.

When the electromagnetic wave with its rapidly

alternating electric field passes through the space occupied by the electric charges it moves them first in one direction and then in the opposite direction as the direction of the field alternates. The vibrations in the wave are therefore "loaded" with these electrons and therefore travel at a lower speed. The electrons themselves will have a characteristic time of vibration within the atom, for if by any means they are displaced from their usual position the attraction of the opposite charges will pull them back again and so set up a vibration.

If the period of the wave is much greater than that of the electron, the simple idea of the loading up of the wave by the electron is quite accurate, but if the period of the wave is very nearly equal to the natural period of one type of electron the simple theory breaks down. When the periods are exactly equal the electron resounds to the wave and vibrates violently, absorbing all the energy of the wave. We thus get an absorption band in the spectrum.

If the period of the wave is a very little greater than that of the electron a more accurate mathematical investigation shows that the speed of the wave is very small, while if the period of the wave is a little smaller than that of the electron the velocity is very large. This is just what is indicated by the greatness of the refractive index on the red side of an absorption band and the smallness on the blue side.

CHAPTER IV

INTERFERENCE OF LIGHT

Definition of Interference.—When two exactly similar sets of waves come up to the same point we might expect that the disturbance due to the two sets together would be that calculated by adding together the disturbances due to each separately, but this is not generally true. At some points there may be no disturbance at all, while at others the disturbance is four times as great. The two sets of waves do not simply add themselves together as if each produced its own effect without any interference from the other, but they interfere with one another. The different phenomena which are due to this interaction between two or more sets of waves are therefore included under the heading of Interference.

Interference of Ripples on Mercury.—In *Radiation*, Chapter I, it is shown how two exactly similar sets of ripples may be started on the surface of mercury, and how these two sets of ripples interfere, producing no effect at all at some points and producing an exaggerated disturbance at others. No disturbance is produced where a crest of one set of waves reaches the point just at the same time as a trough from the other set, for the crest would raise the liquid at the point just as much above the general surface of the liquid as the trough would lower it, and so the liquid is displaced neither up nor down. At other points in the liquid we shall find the two sets of waves reaching them in the same phase, *i.e.* either two troughs or two crests together will reach the points. The liquid at these points will therefore be displaced upwards or downwards by twice the amount due to one set of waves, and since we can show that the energy of the disturbance is proportional to the square of the displacement the energy of the disturbance at these points is four times as great.

No Interference with two Different Sources of Light.—At first sight we might think that light does not exhibit this phenomenon, for if two similar sources of light be placed so as to illuminate the same source we know that the effect is a uniform illumination which is brighter than that produced by either source separately.

This is due, however, to the fact that even though our two sources may be sending out similar waves yet they will be sending them out at different times. Thus one source may be sending out a crest just as the other is sending out a trough, and then a little later they may be sending out new sets of waves which are in phase with one another. We have no means of controlling the emission of waves from two different sources of light, and so we can never get interference between them.

Interference between two Parts of same Set of Waves.—If we divide the waves from a single source of light, however, into two or more parts, these two parts may interfere, for each pulse in the one set of waves will be sent out at exactly the same instant as a corresponding pulse in the other set. Let us then examine the interference between two exactly similar sources of simple waves, S_1 and S_2, Fig. 28, which are sending out waves

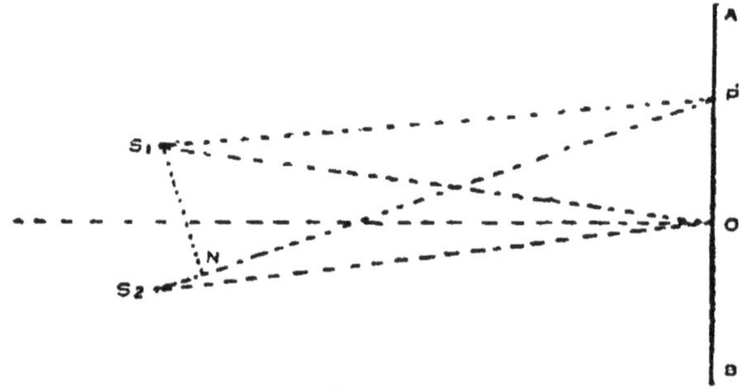

FIG. 28.

to the screen AB. At the point O, which is symmetrical with respect to S_1 and S_2, the waves will arrive in the same phase, for the distance OS_1 is equal to OS_2. They will therefore reinforce one another. If we move outwards from O to P on the screen the distances from P to S_1 and S_2 will no longer be equal, the difference between them being greater the further P is from O. If P is chosen, so that S_2P is just half a wave-length greater than S_1P, a wave from S_2 will reach P at the same time as a wave leaving S_1 half a wave later. We shall therefore have a crest from S_2 reaching P just at the same time as a trough from S_1 and vice versa. At P we shall therefore get no disturbance. If P is removed still further from O a point is reached where the difference between S_1P and S_2P is a whole wave-length. When this is the case a wave from S_2 reaches

P at the same instant as one which starts from S_1 a whole wave later. The two sets of waves will therefore be in step again and will reinforce one another. Moving P still further out we shall reach the point where the difference between S_1P and S_2P is one and a half wave-lengths, and then the waves will be out of step again and will neutralise one another. Further out still the difference will be two wave-lengths: the waves will be in step again and will reinforce one another. This transition from reinforcement to opposition, then reinforcement, then opposition, goes on indefinitely as P recedes from O, and thus there are on the screen a whole series of alternating bands of great and no disturbance. The distance between the bands will obviously depend upon the lengths of the waves which are interfering, for the difference between the two distances S_1P and S_2P has to be a certain number of wave-lengths in order to get a bright band. If we get a mixture of wave-lengths the positions of reinforcement and of opposition will be different for the different wave-lengths and so the bands will become confused. Ordinary white light is either a very irregular disturbance or a mixture of a large number of waves of different lengths. The interference bands of white light are therefore confused and soon disappear as we pass from the centre of the system.

Young's Experiment.—The original experiment of Young was performed, however, with white light, and can be repeated by anyone with a little care. The apparatus is represented in Fig. 29. X is a small screen of lead foil (about 2 inches square) with a small pinhole, P, pierced in it, the pinhole being illuminated by some bright source of light to its left. The light should be well boxed in so that there is very little light in the room. Y is another lead foil screen similar to X, but with two pinholes, A and B. These must be very small and very close together, their distance apart being less than half a millimetre. The easiest way to pierce them is to place the foil on a sheet of glass and pierce with a fine needle. Z is a white screen to receive the light coming from A and B, but it is easier to use a magnifying-glass or eye-piece, E, fixed on a little stand instead of the screen Z. It is convenient to have the three screens about a foot apart.

The waves of light emerging from P fall on the second screen, and two small fractions of them emerge through A and B and act like the two sources S_1 and S_2, in Fig. 28, producing interference bands at the screen Z or the eye-piece E. The bright and dark bands are rather faint if received on a screen, but are very clear when viewed through an eye-piece.

Monochromatic Sources of Light.—Interference bands, or fringes as they are frequently called, are very much

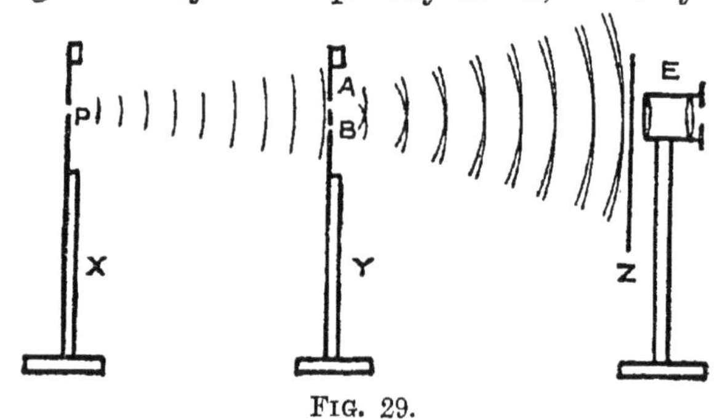

FIG. 29.

clearer and more numerous when a source of light is used which only gives out waves of a certain definite wave-length. Such sources are produced when certain metallic salts are placed in the colourless bunsen flame, the most convenient salts for general use being the salts of sodium such as common salt or carbonate of soda. The sodium flame, which is a bright yellow colour, really gives out waves of two different lengths, but they are so very nearly equal that for a great many purposes they act as waves of a single length.

There are some much easier ways of producing interference bands than by Young's method, and therefore a few of these methods will be described briefly.

Lloyd's Single Mirror Method.—In this method a narrow slit S, Fig. 30, illuminated by a sodium flame, is placed just at the end of a sheet of plate-glass 30 or 40 centimetres long and 3 or 4 centimetres broad. It is parallel to the surface of the glass and a very small distance above it. If an eye-piece, E, is brought up so as

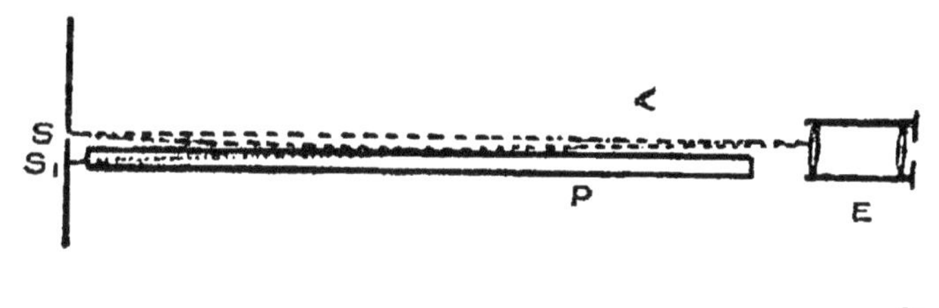

FIG. 30.

to explore the region just above the other end of the plate-glass, the light reaching it will come partly direct from the slit and partly by reflection at the glass. The slit and its image, S′, in the plate-glass will therefore act as two interfering sources and a series of bands parallel to the original slit will be seen.

Fresnel's Mirrors.—In Fresnel's arrangement light

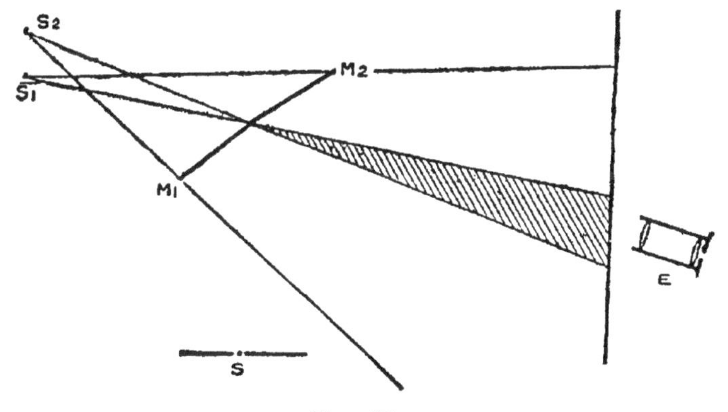

FIG. 31.

from a slit S, Fig. 31, illuminated by a sodium flame is reflected by two mirrors, M_1 and M_2, which are inclined very slightly to one another. The slit is turned so that it is parallel to the line where the two mirrors meet.

The light reflected from M_1 appears to come from S_1, the image of S in M_1, while the light reflected from M_2 appears to come from S_2. Where the two reflected streams overlap they will interfere as if coming from two interfering sources, S_1 and S_2, and so an eye-piece, E, will enable one to observe bands parallel to the original slit. In order to get good bands the inclination of the mirrors must be extremely small so as to get S_1 and S_2 very close together.

Fresnel's Biprism.—In this arrangement the light from a slit S, Fig. 32, passes through a prism of extremely obtuse angle. The light through the upper half

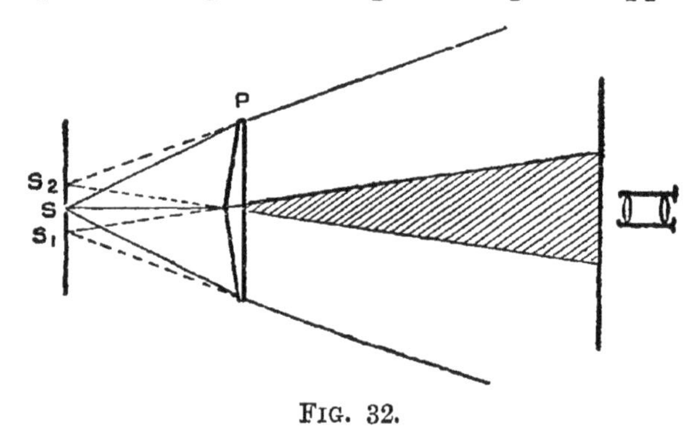

FIG. 32.

of the prism is deflected downwards and therefore appears to come from a source S_2, while the light through the lower half is deflected upwards and so appears to come from S_1. Where the two streams overlap, therefore (the shaded area), there will be interference, and the position of the bands will be just as if S_1 and S_2 were two interfering sources. The bands are most easily received in an eye-piece. By measuring, as can very simply be done, the distance between S_1 and S_2, the distance between two consecutive bright bands in the eye-piece, and the distance between the slit and the eye-piece, the wave-length of the light may be calculated. This method is not very accurate, since the bright bands are not sharp lines whose distance apart can be measured with extreme accuracy, but are bands with a maximum of brightness in the centre gradually shading off to darkness at the middle of the dark bands. The measurement shows that the wave-length is extremely small, however, being a little less than ·00006 centimetres for the sodium flame. This means that the interfering sources must be very close together, $\frac{1}{2}$ mm. or less, in order that the bright bands may be far enough apart to be comfortably observed and measured.

The Colours of Thin Films.—The iridescent colours which are exhibited by thin films of transparent substances are also due to interference. The interference takes place between the stream of light which is reflected at the front surface of the film and that which enters the film and is reflected at the back surface. In Fig. 33, AB represents a ray of light incident on the thin film, magnified greatly in the diagram, at B. Part of it is reflected along BC and the other part refracted along BD. At D part of the ray is reflected along DE, and the other part is transmitted along DG. Of the ray DE part is reflected, and the other part is refracted along EF. The two parallel rays BC and EF interfere when they are brought to a focus at the same place by any lens such as that in the eye. For the reflected ray at B has only to travel from B to H in the air in order to reach the wave-front EH, whilst the ray BDE has to travel the distance BD and DE in the film in order to reach the same position. There is therefore a path difference between the two rays which is greater the greater the thickness of the film and which also increases with the obliquity of the ray. We shall therefore get reinforcement or darkness according as the path difference is a whole number of wave-lengths or an odd number of half wave-lengths. Since the different

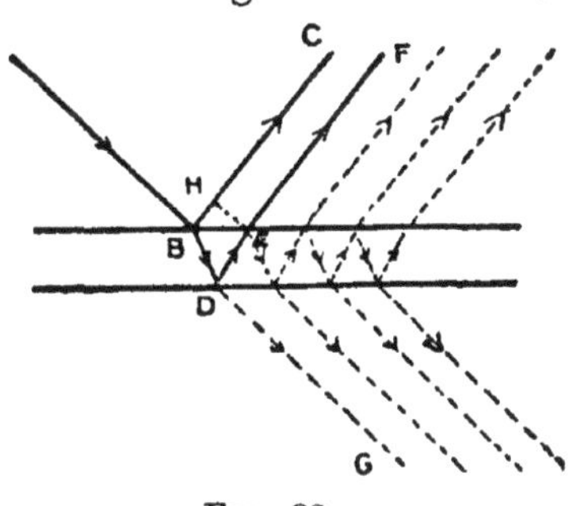

FIG. 33.

colours in the spectrum are merely waves of different lengths the different colours will be reinforced with different thicknesses of film and with different obliquities, and so the well-known coloured effects of soap films and other thin films are produced.

Newton's Rings.—Newton investigated one of the best-known examples of the interference in thin films; and it is ironical that his name should be specially connected with what is so essentially a wave phenomenon. He placed a convex lens of very slight curvature on a piece of plate-glass and viewed it by reflected light. Brightly coloured circular rings were seen immediately surrounding the point of contact between the lens and the plate and gradually shading off into uniform illumination further away from the centre. If monochromatic light, *e.g.* a sodium flame, be used, alternately dark and bright rings are observed which become narrower and closer the further they are from the centre, but which do not shade off into uniform illumination. Thousands of these rings may be observed. Fig. 34 will make their production clear. A ray AB is reflected partly at B and partly at C, and therefore there is a difference of path between the two reflected streams which is approximately twice the distance BC if AB is normal to the surface. If BC is a whole number of half wave-lengths the difference of path will be a whole number of wave-lengths, and we should expect reinforcement, and if it is an odd number of quarter wave-lengths the path difference will be an odd number of half wave-lengths, and we should expect darkness. As the region of equal thickness are rings round the central point of contact we should expect the alternating rings of brightness and darkness. In the centre where the thickness is zero the path difference is zero, and therefore we should expect a bright spot. But the central spot is found to be black, and everywhere

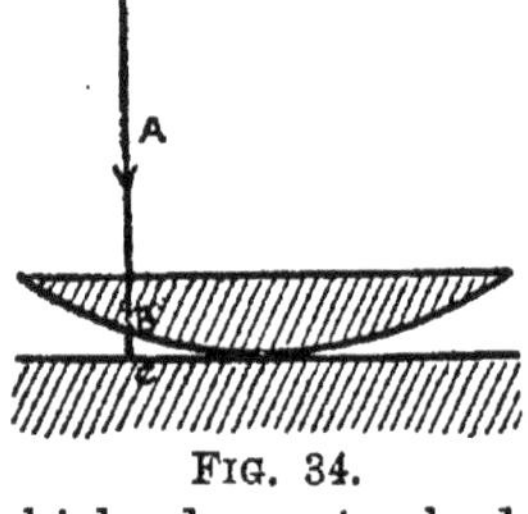

Fig. 34.

where we expect brightness we get darkness, and vice versa.

Change of Phase at Reflection.—The explanation of this was given by Young and is to be found in the fact that the two reflections take place under different conditions. At one surface the reflection takes place at the surface of a rarer medium, while at the other it is at the surface of a denser medium. Those waves which are reflected at the dense medium suffer a change of phase of half a period, which is equivalent to half a wave-length path difference, while those reflected at the rare medium suffer no change at all. We therefore have to add half a wave-length to the calculated path difference in every case and so get a reversal of the expected brightness and darkness.

An idea as to how the reversal of phase may be produced can be gathered from the crude model given in Fig. 6, lower part. Imagine the ball A pulled out and allowed to swing in and strike the ball next to it. It is thereby brought to rest, but the one in front is given an equal velocity. This ball strikes the one in front of it, transmits its own velocity, and is itself brought to rest. This process is repeated until the ball B is reached. The ball B strikes against C, but as C is more massive it does not receive so great a velocity and B bounces back with a velocity in the opposite direction, *i.e.* its phase is reversed. It transmits this velocity back to A, and so the disturbance is reflected with a reversal of phase. This is evidently analogous to the reflection at the surface of a denser medium. If the ball D is pulled out and allowed to swing in, the disturbance will travel from D to C just as it did from A to B. Since B is less massive than C, however, the latter is not brought to rest by the impact, but continues its swing for some little way. It then swings back, strikes the next ball behind it, and so transmits some of the disturbance back again to D. Evidently we get reflection here without change of phase, because after the impact C moves on in the same direction as before.

Young supported his explanation by a pretty experiment. He observed the Newton's rings formed by a lens of crown glass and a plate of flint glass with cassia-oil between them. Now the refractive index of cassia-oil is greater than that of crown glass, but less than that of flint glass, and therefore the reflection at each of the surfaces is at a denser medium. The system of rings under these conditions had a bright centre just as was expected.

Multiple Reflections in a Thin Film.—At first sight we should expect the darkness of the dark rings to be only partial, because the intensity of the beam which is reflected at the lower surface is less than that reflected at the upper. This is due to the two refractions which the first beam suffers in addition to the reflection, and at each of these refractions some light is lost by reflection. Suppose, for example, that at each surface $\frac{1}{10}$ of the light is reflected and $\frac{9}{10}$ transmitted. The beam which is reflected at the first surface will have $\frac{1}{10}$ of the intensity of the incident beam. The refracted beam will have $\frac{9}{10}$ the intensity of the incident beam and $\frac{1}{10}$ of it will be reflected at the second surface, *i.e.* $\frac{1}{10}\times\frac{9}{10}$ of the incident beam. Of this reflected beam $\frac{9}{10}$ will be refracted at the top surface and emerge to interfere with the first reflected beam. Its intensity will therefore be $\frac{9}{10}\times\frac{1}{10}\times\frac{9}{10}$ of the original beam, *i.e.* $\frac{81}{100}\times\frac{1}{10}$. This second beam therefore will be only $\frac{81}{100}$ times as intense as the first beam, and the two could never produce complete darkness. As a matter of fact the centres of the dark rings seem to be perfectly black. This is due to the fact that there will not be two beams emerging parallel, but many. At the point E, Fig. 33, there is again a reflected beam which is partly reflected at the lower surface, and the reflected part is partially transmitted at the upper surface, emerging parallel to the other two rays. This process is repeated indefinitely, but after a number of reflections the intensity becomes so small that we may neglect the remainder of the beams. All these parallel beams will be brought to the same focus by a lens and they will therefore all of them interfere. When we calculate the intensities of all these other beams, just in the same way as we have calculated the first, we find that the sum of the intensities of all the beams from EF to the right is exactly equal to that of BC. It is thus the interference of BC with all the other reflected beams which results in the perfect darkness of the dark rings.

Newton's Rings with White Light.—The study of Newton's rings with light of a single wave-length has shown us that we get the first bright ring where the thickness of the film is a quarter of a wave-length. If we have the mixed waves of white light it is evident that the first bright ring for the violet light will be where the thickness of the air film is a quarter of a wave-length of violet light, and the first bright ring for the red light where the thickness of the film is a quarter of a wave-length of red light. We thus get a ring of the shortest waves (violet) on the inside shading off through the longer waves to the longest waves (red) on the outside. Since the bright rings, even when monochromatic light is used, are not sharply defined but gradually shade off from a maximum of brightness in the centre to darkness in the centre of the dark band, the colours with white light will be rather mixed, but will be practically the ordinary colours of the spectrum. When the film is thicker the colours are more mixed still, and after the first few rings the illumination will appear quite uniform. This doesn't mean that there is no interference, however, it only means that a number of different colours are reinforced at the same place. We have seen that reinforcement occurs if the film is an odd number of quarter wave-lengths thick, *i.e.* $\frac{1}{4}$, $\frac{3}{4}$, $\frac{5}{4}$, $\frac{7}{4}$

. . . wave-lengths. Waves whose lengths are 4 times, $\frac{4}{3}$ times, $\frac{4}{5}$ times, $\frac{4}{7}$ times . . . the thickness of the film will therefore all be reinforced. Suppose that the film is ·000045 cms. thick, then the following waves will be reinforced :

$$4 \times \cdot 000045 \text{ cms.} = \cdot 00018 \text{ cms.}$$
$$\tfrac{4}{3} \times \cdot 000045 \text{ ,, } = \cdot 00006 \text{ ,,}$$
$$\tfrac{4}{5} \times \cdot 000045 \text{ ,, } = \cdot 000036 \text{ ,,}$$
$$\tfrac{4}{7} \times \cdot 000045 \text{ ,, } = \cdot 0000257 \text{ ,,}$$

Of these waves only the ·00006 cms. and ·000036 cms. are in the visible spectrum, the ·00018 being in the infra-red and all the others being in the ultra-violet. Only two colours will be reinforced, therefore, and we shall get a very distinct coloured effect. If the light coming from this thickness of film were received in a spectroscope we should get two bright broad bands in the spectrum, the centres of which corresponded to the two wave-lengths ·000036 cms. and ·00006 cms., and these bands would shade off to darkness on each side of them.

If the film were ten times as thick we should get a great many more waves of the visible spectrum reinforced. The waves reinforced would be :

1st. $4 \times \cdot 00045$ cms. $= \cdot 0018$ cms.
2nd. $\frac{4}{3} \times \cdot 00045$,, $= \cdot 0006$,,
. . . . .
13th. $\frac{4}{25} \times \cdot 00045$,, $= \cdot 000072$,,
. . . . .
30th. $\frac{4}{59} \times \cdot 00045$,, $= \cdot 000031$,,

Waves 1 to 12 will be in the infra-red region and waves 31 onwards will be in the ultra-violet. Waves 13 to 30 will be in the visible spectrum and we shall therefore have eighteen different wave-lengths in the visible spectrum reinforced. The resulting mixture will be quite indistinguishable from white light by the eye, but the spectrum of it will show seventeen dark bands across it corresponding to the waves which are not reinforced and which are in between the reinforced ones in the spectrum. By counting the number of these dark bands across the spectrum we could evidently work backwards and calculate the thickness of the film. This method is sometimes used for measuring the thickness of films.

Interference Spectroscopes.—When Newton's rings formed by sodium light are observed a good way from the centre it is found that they become confused and disappear entirely at about the 500th ring. Afterwards they reappear again, till at about the 1000th ring they are as sharply defined as ever. At about the 1500th ring they disappear again, only to reappear again, and become as clear as ever at the 2000th ring. This is because the sodium light does not consist of waves of a single wave-length, but of waves of two different lengths, one of them ·00005890 cms. and the other one ·00005896 cms. The latter is thus about one-thousandth greater than the other. The thickness of the film to produce the 500th bright ring for the ·00005896 cm. waves would need to be $\frac{999}{4} \times \cdot 00005896$, *i.e.* ·01472 cms. Now this is almost exactly equal to $\frac{1000}{4} \times \cdot 00005890$ cms., *i.e.* to a whole number of half wave-lengths of ·00005890 cms. This means that for the shorter waves there will be opposition and therefore darkness. Thus the bright ring for the longer wave just falls on the dark ring for the shorter wave, and vice versa. There will therefore be uniform illumination. The thickness of the film for the 1000th band of the ·00005896 cm. waves will be $\frac{1999}{4} \times \cdot 00005896$ and this is almost exactly equal to $\frac{2001}{4} \times \cdot 00005890$ cms. The latter number is the thickness of the film for the 1001st ring of the ·00005890 cm. waves, so that the 1000th bright ring of the larger waves just falls on the 1001st bright ring of the shorter ones and the rings appear as sharply defined as ever.

By observing the bands of a very high order we have thus been able to glean some information as to the constitution of a line in the spectrum, and have found it to be double. This observation of bands of a high order is the main principle in interference spectroscopes. Some of these are too complicated for description in this book, but the principle of one of them is explained here.

Michelson's Interferometer.—The arrangement devised by Michelson is shown in plan in Fig. 35. A and C are

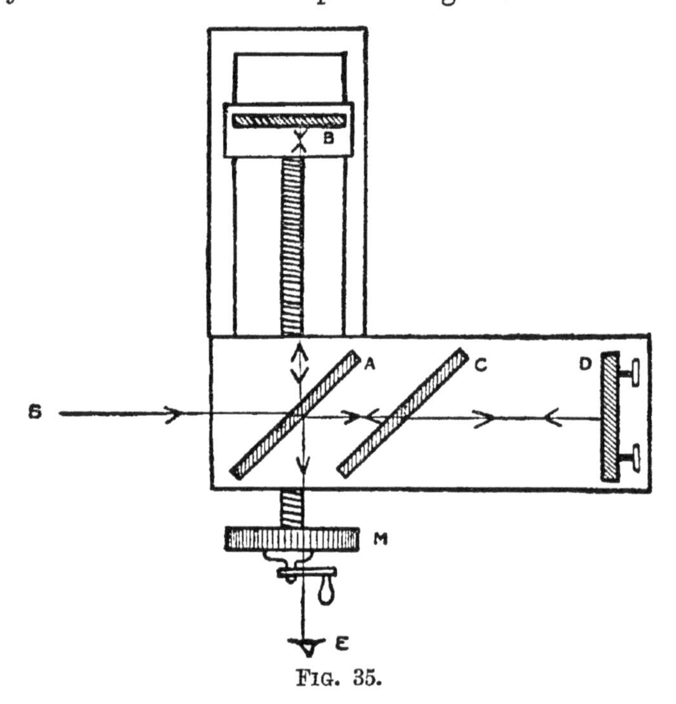

Fig. 35.

two vertical plates of perfectly plane glass of exactly equal thickness placed parallel to one another. B and D are two perfectly plane vertical mirrors silvered on their front surfaces. D has adjusting screws at its back so that it may be turned horizontally or vertically through a small angle. B is fixed on a carriage so that it may be moved perpendicular to itself by means of a fine screw, which is turned by a milled head, M. A and C are at 45° to B and D, the latter being perpendicular to one another. The right-hand surface of A is usually very lightly silvered so that about half of any light coming up to it is reflected and about half transmitted. A ray of light from a monochromatic source, S (*e.g.* a sodium flame), can enter an eye placed at E by either of two paths. The first part is reflected at the back surface of A, travels to B and back again through A to the eye. The second part is transmitted through A and C, is reflected at D, transmitted back through C and reflected at the half-silvered surface of A. It will be noticed that the first ray traverses the thickness of A three times, and the second traverses A once and C twice. If C and A are of exactly equal thickness the two rays will traverse exactly equal thicknesses of glass, and therefore if the air distances are the same there will be no path difference between the two parts. By moving B by means of the screw, however, any desired path difference between the two parts can be produced and so we may get reinforcement or destruction of the light almost as many times as we choose.

If the eye observes a little obliquely, so that the rays entering it are no longer exactly normal to B and D, the path of one of the rays is increased, while that of the other is decreased, and if the obliquity is sufficient to cause an extra wave-length difference in path we shall have reached the next reinforcement. Since the points of equal obliquity lie on a circle round the central line perpendicular to B, a system of circular bands will be seen, though when the path difference is very great and the centre of the system right off the field of view the circles may be so large that the portions of them in the field of view are nearly straight lines. It is interesting to notice the appearance and disappearance of the bands when a sodium flame is used, due to the dual character of the waves. Twenty or thirty different disappearances and reappearances have been observed, corresponding to a difference of path of twenty or thirty thousand wave-lengths between the two parts of the ray.

Light Waves as Standards of Length.—One of the most interesting uses to which the interferometer has been put has been to determine with most extraordinary accuracy the wave-lengths of the three different colours which are emitted by salts of cadmium when placed in a bunsen flame. The lengths of the waves emitted by a luminous metallic vapour seem to be extremely constant, and so the ultimate idea in this determination was to use these wave-lengths as standards of length. The method adopted was in principle to count the number of bands which cross a fixed line in the field of view as the movable mirror was moved back a certain distance. The number of bands crossing will be the number of wave-lengths in twice this distance. An idea of the accuracy of the work may be gathered from the following three values of the number of cadmium red wave-lengths in one metre, obtained two of them by Michelson and one of them by Benoit:

1553162·7. 1553164·3. 1553163·6.

The difference between the extreme values is just about one in a million, and we therefore have the length of the standard metre expressed to this accuracy in terms of a natural invariable unit. Probably the number of copies of the standard metre are too numerous for their loss or destruction to be at all likely, but if no accurate copy of the standard were attainable it could be reproduced by making use of this determination.

CHAPTER V

DIFFRACTION

Definition of Diffraction.—We have seen in the chapter on the rectilinear propagation of light that light bends into the geometrical shadow of an obstacle to a small extent, the completeness of the shadow being due to the extreme shortness of the wave. This simple bending of the light round obstacles is called "diffraction," but we usually include under the term a number of interference effects which accompany it. As was mentioned earlier in the book, Grimaldi first observed diffraction and described his experiments in 1665. He admitted sunlight into a darkened room through a very small aperture and noticed that if an obstacle were placed in the path of the rays coming from this aperture the edges of the shadow were bordered by a rainbow-tinted fringe, and if a very small object was used the fringes appeared in the shadow of the obstacle itself. Grimaldi, Newton, and Young all tried their hands at explaining these fringes, but Fresnel gave the first true explanation, using the method of dividing the wave-front up into zones, as he had done in explaining the rectilinear propagation. In estimating the illumination at any point he observed how many of the zones were intercepted by the obstacle and subtracted the effect of these from the effect of the whole wave-front.

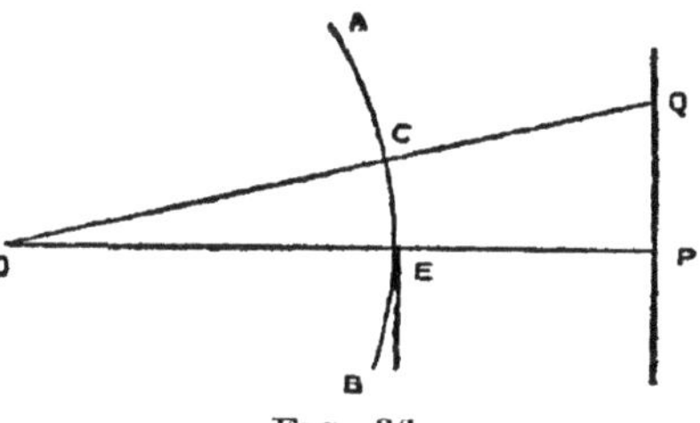

FIG. 36.

Diffraction at a Straight Edge.—In Fig. 36, O represents the illuminated aperture and E a straight edge perpendicular to the plane of the paper, casting a shadow on to the screen, S. AB represents a wave-front arriving at the straight edge, E, at a certain instant. If we consider the illumination at P, the edge of the geometrical shadow, we see that the edge E cuts straight across the centre of the system of zones, because PE is perpendicular to the wave-front and therefore E is the centre of the zones. Since each element is halved the illumination at P will be half that if the obstacle were absent, *i.e.* quarter that due to the first zone. For a point Q the point C will be the centre of the zones, and there will be a complete half of the system of zones above C. Below C we shall have in addition the other half of a few of the more central zones.

First suppose CE is just the width of the first zone, then the illumination at Q is due to half of the whole system of zones plus half of the first zone. The intensity will therefore be three-quarters that due to the first zone, or three times as much as at P. Now suppose CE is the width of two zones, then the illumination at Q will be due to half of the whole system plus half of the first two zones. The first two zones practically neutralise one another and therefore the illumination will be equal to that due to half of the whole system, *i.e.* will be the same as at P. Moving Q still further out until CE is the width of three zones, two of these three neutralise one another, leaving the third effective, so that we shall again get the intensity equal to that due to half the system plus half of one zone, *i.e.* three-quarters of that due to the first zone. Moving Q further and further out it passes through a series of maxima and minima of brightness which get less and less marked and shade off into uniform illumination.

Since the size of the zone depends upon the wave-length the maxima of brightness will be in different positions for different wave-lengths, *i.e.* for different colours, and the bright and dull bands therefore will be coloured.

Within the geometrical shadow the edge cuts off more than half of the whole system, the number of half zones left being smaller as the point is further within the shadow. The brightness of the illumination therefore falls off continuously but rapidly as we enter the shadow. The edge of the shadow is thus not distinctly marked on the dark side and has a number of coloured fringes on the bright side.

Diffraction at a Narrow Slit.—The fringes due to a narrow slit may be seen very easily by placing a narrow slit just in front of a bright light and viewing it through another narrow slit held parallel to it, close to the eye, at a distance of a yard or two. Serviceable slits may be made by ruling a line on silvered glass with the point of a sharp knife, or even by ruling a line with a sharp knife in lead foil placed on a piece of glass. The slits, however, must be straight and narrow.

The diffraction pattern will be seen to consist of a central bright band bordered by alternately bright and dark coloured bands which shade off rapidly in intensity as they are farther from the centre. The following simple treatment will be sufficient to explain their production. Let AB, in Fig. 37, represent the slit (magnified) close to the eye, and S the screen on which the light is received, *i.e.* in this case the retina of the eye. At the point P the light coming from all parts of AB has all to come practically the same distance and therefore will be in the same phase and produce brightness. At a point Q, however, the light coming from different parts of the slit has to come different distances. Suppose Q is removed such a distance from P that BQ is just half a wave-length longer than CQ, and CQ half a wave-length longer than AQ. Then for each point in the upper half of the slit we can find a point in the lower half which is just half a wave-length further away from Q. The two halves of the slit will therefore mutually destroy one another's effect at Q, producing a dark band. Moving further out we arrive at a point R, such that BR is one and a half wave-lengths longer than AR. If we now divide the slit into three equal parts by D and E, BR will be half a wave-length longer than DR, DR half a wave-length longer than ER, and ER half a wave-length longer than AR. Two of these portions of the slit will mutually destroy one another's effect, leaving the third portion effective. We shall therefore have a bright band at R of about one-third the brightness of the central band. Further out still, the difference between the distances from A and B becomes two wave-lengths, and the slit can be divided into two mutually

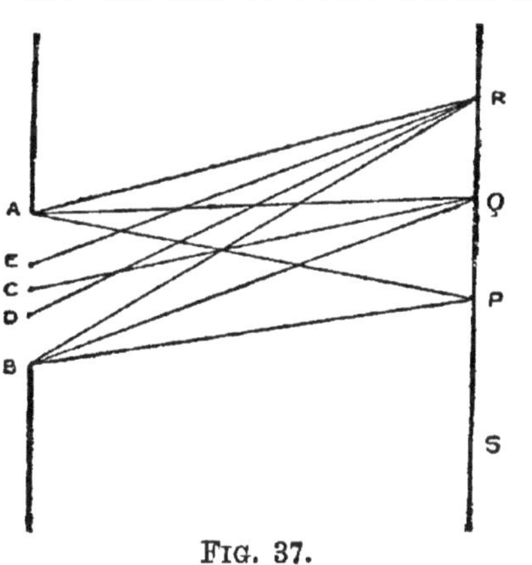

Fig. 37.

destructive pairs of quarters, and therefore another dark band results. Further out still we reach a point where the distances from A and B differ by two and a half wave-lengths, and then the slit may be divided into fifths, of which two pairs are mutually destructive, leaving the fifth one effective. A bright band of about one-fifth the intensity of the central one will therefore be produced. Evidently this process will continue until the intensity of the bands becomes too feeble for them to be observed. Since the distance of the bands from the centre depends upon the wave-length, they will be coloured unless monochromatic light is used. It is evident, too, that the narrower the slit used the further must we move out from the centre of the screen in order to produce a certain difference between the distances from the two edges of the slit. Thus the narrower the slit is the wider apart are the bands.

Diffraction by a Narrow Wire.—This case too is very readily observed. Stretch a very fine wire or fine fibre of any sort across a small convex lens, and, holding the other surface against the eye, look through the lens at a slit placed just in front of a bright light, taking care to keep the slit and the fibre parallel. On each side of the shadow the ordinary fringes due to a straight edge are observed, but within the shadow another set of bright bands are distinctly seen which are easily distinguished from the others by being more distinct, narrower, and equally spaced. This case may be looked upon as complementary to the narrow slit, for here we have that portion of the wave stopped which was transmitted by the slit, and vice versa. The treatment too is very similar. Let O, in Fig. 38, represent the slit with the bright light behind it, AB the width (very much magnified) of the wire, and M and P points on the screen or retina. The illumination at P is due to the whole wave-front minus the zones, which are intercepted by AB. At the point M in the centre of the shadow of the wire we get a number of central zones cut off symmetrically by the wire, and therefore the intensity, like that along the axis of a circular obstacle, is nearly the same as if no obstacle existed. At any other point P, however, the zones are not cut off symmetrically. Beside those which are intercepted entirely there are some which are cut off on one side only. If there are an even number of these latter they will neutralise one another's effect, and the intensity will be the same as at the centre, M.

If an odd number of these half zones are cut off one extra half zone will be effective. This will be the zone immediately within the first complete zone transmitted round the obstacle. When considering the circular obstacle (on p. 19) we saw that the effect of any particular zone and all those outside it is equal to half this first

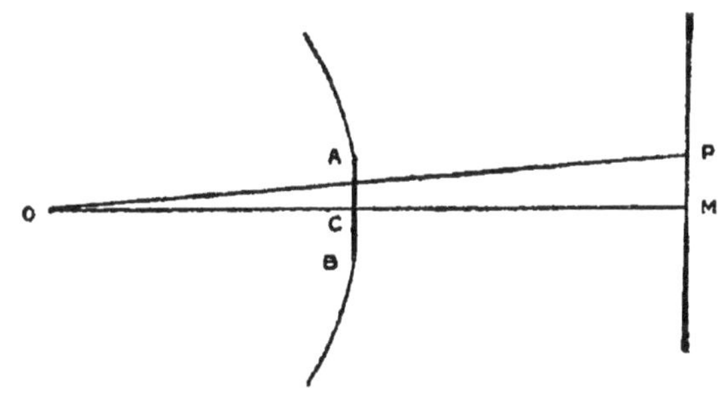

FIG. 38.

zone. The total effect in this case, therefore, will be one-half of the first complete zone plus one-half of the zone immediately within it. The result will be darkness, for these two half zones will mutually destroy one another's effect. As the point P moves away from M on either side it will evidently move successively through points where the wire cuts off odd and even numbers of half zones, and hence we get light and dark bands parallel to the slit and the wire.

Diffraction by a Circular Aperture.—The diffraction caused by a circular aperture cannot be observed as readily as one might expect, owing to the difficulty in producing a truly circular small aperture. If the aperture is broader by a wave-length or two in one direction than in the perpendicular direction the diffraction bands are quite altered. With a small perfectly circular aperture the bands are very easily seen. A pinhole is placed just in front of a source of light, and is viewed from a little distance with the small circular aperture placed just in front of the eye. The pinhole is seen surrounded by a series of coloured rings.

The circular aperture may be treated approximately in the same way as the slit, thus accounting for the maxima and minima of brightness as we move away from the central point. The strict investigation is a little beyond this book.

The chief interest of this case lies in its application to optical instruments. For example, when a telescope views a star there are just exactly the conditions for producing these fringes due to a circular aperture, for the star is a point source of light, and the objective of the telescope is itself the circular aperture. True, it is a large aperture, and therefore the rings are very small; but if two stars are very close together in the telescope the rings due to them may become mixed, and the stars therefore cannot be distinguished or "separated" by the telescope. The larger the objective of the telescope the smaller will be the diffraction rings, and therefore the more able will the instrument be to distinguish objects which are extremely close together. This is the main object in having extremely large telescopes. At the same time a much larger amount of light can enter the telescope through the large objective, and so feebly luminous bodies may be seen which would be invisible with a smaller aperture.

Diffraction due to a Circular Obstacle.—Just as narrow wire acts as the complement to a narrow slit and produces similar bands, so a circular obstacle acts as the complement of the circular aperture and produces similar circular bands. Surrounding the bright patch in the centre of the shadow, the reason for which we have already discovered, there are alternating bright and dark rings, which are coloured when mixed light is used.

The simplest method of observing them is to look through a small pinhole in a card, placed just in front of the eye, at the sky or any other bright object. Apparently floating about in the sky, one sees a number of small sets of rings very like the rings due to a circular aperture. They are really the diffraction patterns on the retina due to small circular particles floating about in the lens of the eye.

The strict investigation of the circular obstacle, like that of the circular aperture, is beyond the scope of this book.

Coronas.—If instead of a single circular obstacle we have a large number of equal ones, then they will all reinforce the light sent in certain directions and destroy it sent in other directions, the directions being the same for all the particles because they are all the same size. We shall therefore get the same pattern as for a single obstacle, only much more intense. A very simple way of observing it is to breathe on a piece of clear glass and then look through it at a source of light some distance away. The light will be seen to be surrounded by coloured rings, but they are not very brightly coloured, because the particles of moisture on the glass are not all of the same size. If instead of breathing on the glass lycopodium is sprinkled on it the rings are very striking, because the lycopodium particles are all equal little spheres. The rings round the moon and sun in hazy weather, or round a street lamp in a fog, are due to this same cause, a thin cloud of nearly equal particles of moisture coming in between the observer and the source of light. These rings are called coronas. They closely surround the source of light, and must not be confused with the large rings or halos formed at some distance away, which are due to refraction, not to diffraction.

Diffraction due to two Parallel Slits.—The arrangement to observe this case would be simply a slight variation from Young's experiment. The pinholes in Young's experiment are replaced by parallel slits, otherwise the experiments are exactly similar. It is evident, therefore, that we shall get a series of dark and light bands parallel to the slits. Suppose A and B, in Fig. 39, are the two slits placed close together, and that the rays of light are coming up in the direction of the arrows from the slit some distance away.

If the point P is chosen on the screen, so that AP = BP, the wavelets emerging from A and B will reach P in the same phase, and will therefore reinforce one another and form a bright band. If Q is chosen, so that AQ is just half a wave-length longer than BQ, then the wavelets emerging from A and B will reach Q in opposite phases, and therefore we shall get darkness. As we move out from P to Q the two wavelets will be getting more and more out of step until this opposition of phase is reached. We shall therefore get a gradual shading off from the maximum brightness at P to the complete darkness at

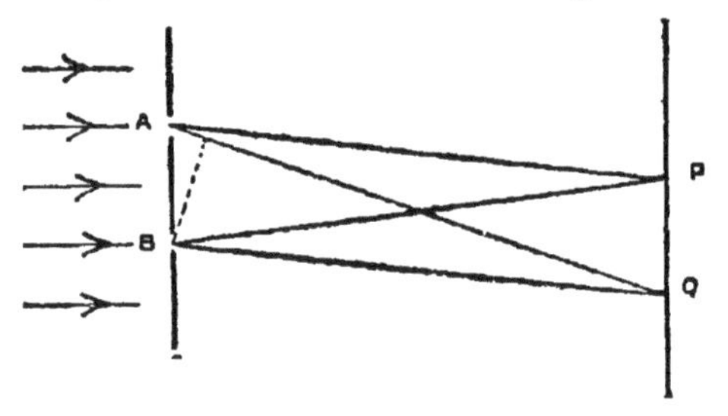

FIG. 39.

Q. This gradual shading off is characteristic of all the bands produced by interference or diffraction which we have considered up to the present. Now suppose we have four slits equally spaced instead of two. We shall get agreement of phase at P for all four slits and opposition of phase at Q for the two pairs of slits, 1 and 2, and 3 and 4. But at a point halfway between P and Q we shall also get opposition of phase between the two pairs, 1 and 3, and 2 and 4, for they are twice as far apart as the first two pairs. We shall therefore get darkness at Q and at a point midway between P and Q. The bright band will therefore be narrower, and the dark band broader, than with two slits. If we represent graphically the brightness at different distances from P we shall get a curve something like Fig. 40, where the thick solid line represents the brightness with two slits, and the dotted line with four slits. We see that there are two small subsidiary maxima with the four slits, and these of course appear as faint bands in between the bright main bands. If we have eight equidistant slits we can divide them into pairs, 1 and 5, 2 and 6, 3 and 7, 4 and 8, which are four times as far apart as AB, and these will therefore produce darkness at a point only a quarter as far as Q from P. The intensity of the central band therefore falls off to zero very rapidly, as shown in the thin solid line in Fig. 40. It will be noticed that we have six subsidiary maxima in between the main maxima, which, if visible at all, will appear as very faint

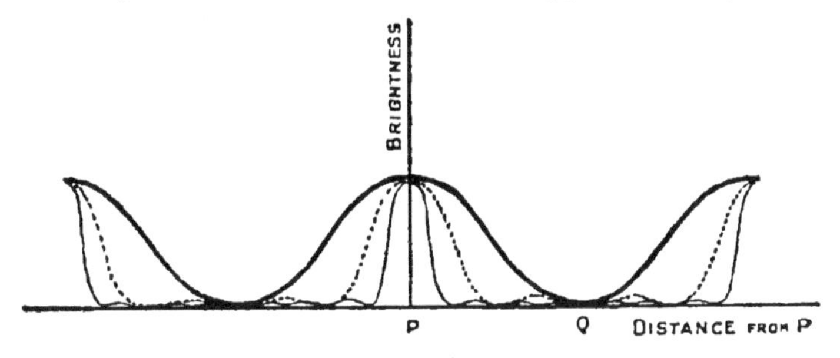

Fig. 40.

bands in between the bright ones. Every increase in the number of equidistant slits leads to a narrowing of the bright band and a broadening of the dark space in between, until when the number of slits is very large the bright bands are sharp narrow lines separated by broad dark spaces. The distances of these lines from the centre will depend upon the wave-length of the light, being greater for the greater wave-lengths. Each colour will therefore produce a sharp narrow line, and we shall get a pure spectrum without any overlapping of the colours.

The Diffraction Grating.—The diffraction grating is merely a large number of parallel equidistant slits. A rough grating is provided by a piece of fine wire gauze, or even by a thin plain silk handkerchief. Looking through the grating at an illuminated slit, or a gas flame turned edgewise, we see on each side of the slit or flame coloured images which are violet on the inside and red on the outside. To be of much use the gratings must be very much finer than these, for the larger the number of slits the finer are the bright lines, and the closer together they are the greater is the distance between the lines, and therefore the greater is the dispersion in the spectrum. The best gratings are made by ruling fine lines with a diamond point on glass or polished speculum. In the glass gratings the spaces between the scratches act as slits, the scratches themselves being opaque. In the speculum gratings the spaces between the scratches reflect the light instead of transmitting it, the scratches themselves being rough and therefore scattering the light. The action is practically the same as for the glass grating. The ruling of gratings has been developed into a very fine art by Rowland, who has succeeded in ruling speculum gratings 6 inches across with 20,000 lines to the inch. The production of a really good grating is a very tedious and laborious business, and is a good deal a matter of chance even then. The actual ruling of a 6-inch grating with 20,000 lines to the inch takes about five days and nights after everything has been prepared, and there is always the possibility that the diamond point may chip in the middle of the ruling and so spoil the whole grating. Any serious alteration of the temperature during this time will also spoil the grating, for it will cause the speculum to expand or contract and so alter the spacing of the lines. Practically all gratings have some imperfections in the ruling, and if these are serious they sometimes cause the appearance of faint extra images or "ghosts."

Until the advent of the modern interference spectroscopes the diffraction grating was by far the most accurate instrument for measuring the wave-length of any monochromatic light. The calculation is very simple. Let A and B, Fig. 41, be two adjacent slits of the grating, and let S be the screen or the retina on which the light is received. The screen is placed at the principal focus of the lens, L, so that parallel rays coming from the grating will be brought to a focus on it. If the screen is the retina, L is the lens of the eye. If a plane wave comes up to the grating in the direction of the arrows the wavelets transmitted straight on will be brought to a focus at P, and since there will be no difference of path between them P will be the central bright line.

Moving out from P we shall reach the first bright line,

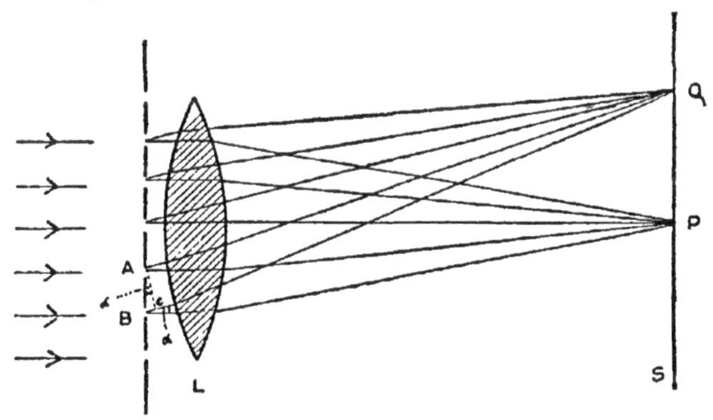

FIG. 41.

Q, where the difference of path between two successive wavelets is equal to a whole wave-length. Drawing AC perpendicular to the rays going to Q, BC is the difference of path. If we call α the angle between the rays going to P and those going to Q, the angle BAC is also equal to α and $\frac{BC}{AB}$ is sin α. The wave-length therefore is AB sin α. The distance AB is known from the number of lines which have been ruled to the inch on the grating, and therefore by measuring the angle α we may measure the wave-length.

CHAPTER VI

POLARISATION

THE properties which we have described up to the present, Reflection, Refraction, Interference, Diffraction, are properties which are possessed by all kinds of wave motion. We now come to some phenomena which depend upon the character of the luminous waves. In a wave of sound the particles of the air or of any other medium through which the wave is travelling vibrate backwards and forwards along the line along which the wave travels. Such waves, which are called longitudinal waves, can obviously possess no onesidedness, for the vibrations must appear exactly the same from whatever side they are observed. But we shall see now that rays of light are produced which do possess a definite onesidedness. The only way in which we can explain this is to suppose that the vibrations in the wave are perpendicular to the direction in which the wave travels. Such waves are called transverse waves. A rough model, shown in Fig. 42, will make the conception clear. A and B are two fixed screens with slits in them and the slits are both placed vertical in the first case. LF is a cord which is fixed at F and which passes through the two slits. The loose end, L, of the cord is jerked about in all directions so as to send a series of waves along the cord. Since the cord can only move up and down in the first slit, only the vertical vibrations will emerge on the other side of A, and since B has also a vertical slit, these vibrations will pass freely through it as well. In the lower half of the diagram the screen B has been turned so that the slit is horizontal and will therefore transmit only horizontal vibrations. None of the waves will therefore be transmitted beyond it. Thus it all depends upon how B is turned as to whether it is transparent or opaque to the waves. Waves like those between A and B, in which the vibrations are all in one direction (in this case vertical), are said to be plane polarised, the plane of polarisation being the plane to which the vibrations are all perpendicular. In this case the plane of polarisation is the horizontal plane.

The polarisation of light was first observed by Bartholinus in 1670. He found that a ray of light when passed through a crystal of Iceland spar was split up into two separate rays of equal brightness. Thus on looking through the crystal everything appears doubled. On passing the rays through a second crystal they are again doubled, but now the rays are usually of unequal

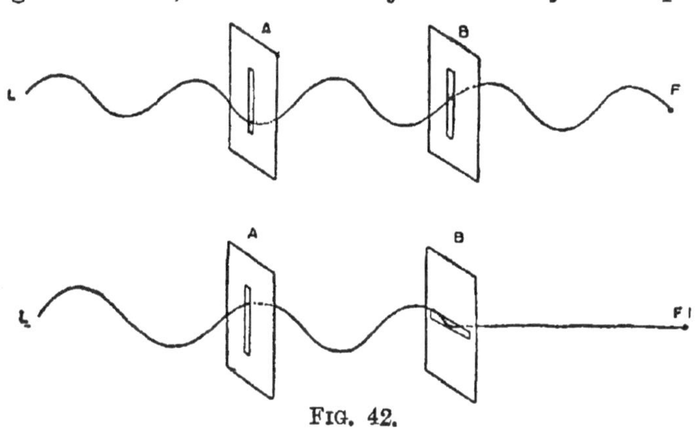

FIG. 42.

brightness, and the relative brightness varies as the second crystal is rotated. In two positions of the second crystal two of the rays disappear completely, so that there is in these positions no subdivision of the rays by the second crystal. Perhaps the easiest way in which to observe this is to place one crystal on some printed paper and note that the print appears doubled. Holding the second crystal in the hand and looking through it, at the first the printing appears quadrupled, but two of the images are fainter than the other two. On twisting the crystal round in the hand, while looking through it, one pair of images of the print grows brighter, and the other fainter, until the latter disappears altogether. On continuing the rotation they reappear again and grow brighter, the other pair gradually growing fainter and disappearing in their turn when the crystal has been turned through 90° from the first disappearance. This process is repeated for every 90° rotation.

This experiment is sufficient to show the existence of light which is onesided about its line of propagation, but Huygens was quite unable to suggest any explanation of it, and it remained an isolated, unexplained fact until 1814, when Fresnel introduced the idea of transverse vibrations and so gave the key to the whole subject of polarisation.

Crystals of Iceland spar, in common with other crystals, have different properties in different directions in the crystal. For instance, the crystal is stronger in one direction than it is in a perpendicular direction, just as wood is stronger along the grain than across it. The result, in the case of wood, is that vibrations along the grain of the wood travel much faster than vibrations across it, and therefore sound travels much faster along the grain than across it. In an analogous way the light vibrations in one direction in the crystal will travel faster than those in a perpendicular direction. Now any form of vibration in a plane can be considered as consisting of two vibrations along two lines at right angles to each other. Let us choose these two lines so that they are in the two directions in the crystal corresponding to along the grain and across the grain in wood, and let us think of the two component vibrations as they enter the crystal. One will travel faster than the other and will therefore have a smaller refractive index, and the two components will be separated. We shall therefore have two polarised rays in place of one unpolarised ray. If the original ray be quite unpolarised, *i.e.* the vibrations in it are equally intense in all directions, the two components are of equal intensity, but when we come to the second crystal we start with a plane polarised ray. Suppose AB, in Fig. 43, represents the extent of the vibrations of a particle in the polarised ray, and suppose OS and OT are the two directions of the second crystal analogous to along the grain and across the grain.

Then drawing AL and BM perpendicular to OT and AB, and PQ perpendicular to OS, LM and PQ will represent the extent of the vibration of the two component vibrations in these directions which are equivalent to the vibration AB. These are the two components which become separated by the second crystal and so redouble the ray into four plane polarised rays. In the diagram it is evident that LM is much smaller than PQ, so that the vibrations are smaller and the light feebler in this particular ray. Turning the second crystal round is equivalent to turning the two lines OS and OT round in the figure. Suppose we turn them in the direction of the arrows, then as OS approaches OA, OP becomes larger and OL smaller until, when OS lies along OA, OP is equal to OA and OL is zero. In this position, therefore, all the light is concentrated into the one beam, OA, and there is no doubling of the ray by the second crystal. Turning through another right angle OT will lie on OB, and OS will be perpendicular. OP will now be zero, and OL will be equal to OB, so that the whole of the light is concentrated into the other beam.

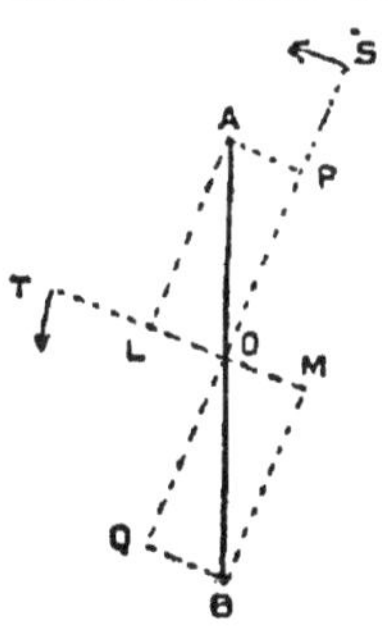

FIG. 43.

This transition from one beam to the other will evidently recur at each rotation of a right angle. If we could get rid of one of the beams coming from a crystal of Iceland spar or some similar crystal we should have a convenient way of producing perfectly plane polarised light. One crystal, the tourmaline, has the curious property of doing this for itself, for it absorbs one of the rays so strongly that if a proper thickness is chosen that ray is practically entirely absorbed. The other ray is rather strongly absorbed, however, so that there is necessarily a great loss of light when the tourmaline is used. Since Iceland spar occurs in crystals of large size and great transparency, it is much more suitable for constructing polarising prisms. Nicol eliminated the ray with the greatest refractive index by means of total reflection. He cut a rhomb of the spar in two along the diagonal plane AB, Fig. 44, polished the cut surfaces and cemented them together again with Canada balsam. Canada balsam has a refractive index in between that of the two rays in the spar, and therefore when the ray of greater index comes up to the balsam surface it is similar to when a ray in glass impinges on the air surface. The prism was therefore cut to such a length that the angle of incidence of this ray on the balsam surface was greater than the critical angle and therefore the ray was totally reflected. The other ray passes from a medium of smaller refractive index to one of very slightly larger, and therefore total reflection is impossible, in fact the difference in the refractive index is so extremely small that the ray is transmitted with scarcely any reflection at all. These polarising rhombs are called Nicol prisms, and are quite the simplest and most effective means of obtaining plane polarised light. If in place of A and B, in Fig. 42, we place a pair of Nicol prisms, the analogy is complete. The Nicol A which separates the plane polarised component from the original irregular vibration is usually called the polariser; the second one, B, which is transparent or opaque to the polarised beam, according to the direction in which it is turned, is usually called the analyser. If the beam of light coming to a Nicol is only partially polarised, the plane polarised part only can be completely extinguished, so that we shall never get complete extinction of the light, but we shall get a maximum and minimum of brightness as the Nicol is turned.

A

B

FIG. 44.

Polarisation by Reflection.—When light suffers reflection at the surface of water or glass, or in fact of any transparent substance, it is found to be partially polarised. At a certain angle of incidence, which depends upon the particular substance used, the polarisation of the reflected beam is most complete. This angle is called the polarising angle. For some substances, *e.g.* glass, the polarisation is practically complete at the polarising angle. The phenomenon can be readily shown by a simple arrangement devised by R. W. Wood and shown in diagram in Fig. 45. The reflectors A and B are two pieces of plate-glass which are painted on the back with black paint or asphalt varnish. This ensures us only getting the reflected beam from the front surface, the beam which is transmitted at the front surface being absorbed by the black backing. The mirror A is fixed at 57° to the horizontal on some convenient stand, while the mirror B is fixed at the same angle, but on a vertical axle which can be rotated by a belt and a pulley, or by any other convenient means. A cylindrical ring of parchment paper or of some other translucent material surrounds the revolving plate and receives the light reflected from it. A bright source of light is placed in the direction S so that a ray reflected from A falls vertically on B and is reflected by it on to the parchment, making a patch of light there. If B is turned so that it is parallel to A a fairly bright patch of light is seen on the parchment, showing that the light coming up to it is quite copiously reflected. If it is now turned through 90° in either direction from this position the patch of light disappears, and if turned through a further 90° it reappears as brightly as in the first position. In the intermediate positions the light is less copiously reflected. If B is rapidly revolved we shall see a ring of light on the parchment with two maxima and two

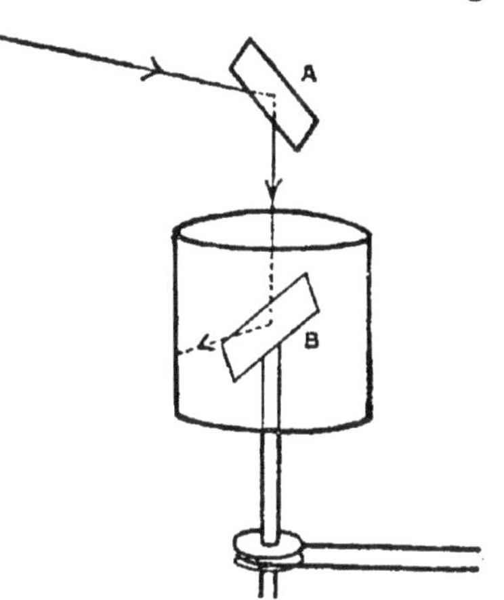

FIG. 45.

minima of brightness corresponding to the positions in which the light is most and least freely reflected. It is quite simple to imagine why light is polarised by reflection, for it is quite conceivable that vibrations parallel to the plane of the glass are less likely to penetrate than those perpendicular to the glass. More of the parallel vibrations will therefore be reflected, and more of the perpendicular ones transmitted.

33°

FIG. 46.

Polarisation by Refraction.—We should therefore expect the refracted beam to be polarised as well, and this is found to be the case. The refracted beam is only partially polarised, however. If we make it pass through a succession of plates all placed at the polarising angle, each plate will reflect more of the parallel vibrations, so that after passing through a number of plates the transmitted beam is practically completely polarised in the perpendicular direction. Eight or ten plates of glass will almost completely polarise the transmitted beam. A pile of strips of glass fixed in a tube, as in Fig. 46, will therefore make an effective substitute for a Nicol prism.

CHAPTER VII

THE ELECTROMAGNETIC NATURE OF LIGHT WAVES

THE evidence given in the preceding chapters shows that light consists of transverse waves of extremely short wave-length. We now wish to know the nature of these waves. In *Radiation*, Chapter I, an attempt is made to give an idea of what an electromagnetic wave is, and in this chapter we will just mention the main evidence on which we suppose light to consist of such waves.

Firstly, light is affected by a magnetic field in several ways.

The Faraday Effect.—The first effect we will consider

is called the Faraday effect. Faraday used an arrangement similar to that in Fig. 47. A is a block of a transparent solid substance or is a glass tank containing a transparent liquid, and N and S are the poles of a powerful electromagnet which are pierced so that it is possible to look straight through them and through the transparent substance. A pair of Nicol prisms, P and A*n*, are placed in the pierced poles, one of them to act as a polariser and the other as an analyser. A source of light is placed in front of P so that it can be observed through the poles, and the analyser A*n* is rotated until it extinguishes the light completely. If the current in

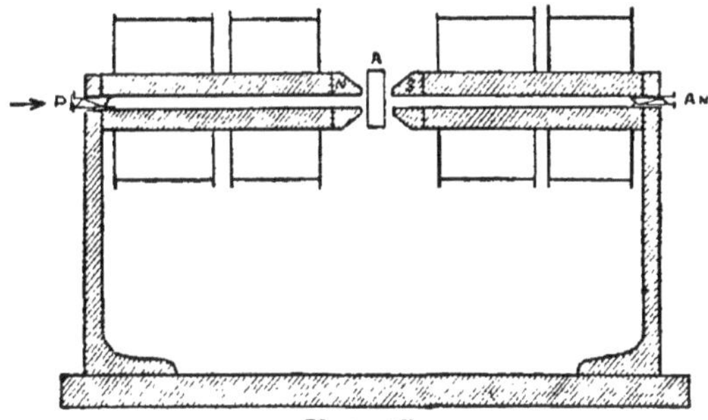

FIG. 47.

the electromagnet is now switched on, the field of view through A*n* will suddenly brighten, so that the rays which were completely stopped by A*n* before any magnetic field was established across A are now able, at any rate partially, to get through the Nicol. If the Nicol is rotated a new position can be found where the light is again completely cut off, so that the light is still polarised, but the plane of polarisation has been turned. This phenomenon of rotation of the plane of polarisation is called rotatory polarisation.

An idea of the way in which this rotation is brought about may be obtained by means of the idea that was introduced in order to attempt to explain dispersion. When an electric wave travels through a transparent substance the small electric charges or electrons in the atoms of the substance are caused to vibrate backwards and forwards with the electric vibrations in the wave. In the arrangement to show the Faraday effect the vibrations are plane polarised, *i.e.* are in one direction only, and since the light is travelling along the lines of force of the magnet the magnetic force is always perpendicular to the direction of vibration. In Fig. 48, BOA represents the direction and extent of the vibrations of the electrons in the wave, and the magnetic field is downwards through the plane of the paper, *i.e.* we may imagine the N pole above the plane of the paper and the S pole below it. Now, whenever an electric charge is moving in a magnetic field it experiences a force which is perpendicular both to the magnetic field and to the direction in which it is moving. The result is that as the small negative charge, the electron, is moving up from O to A it experiences a force to the right which bends its path out so that it moves along the dotted line OLP. As the electron begins to move back in the opposite direction the force introduced by the magnetic field will be reversed, and so instead of continuing to move away to the right the electron will be pulled into the left again and travels back to O along the dotted line PMO. Continuing beyond O to its extreme position on the opposite side the electron will be pulled still further to the left and will travel along ORQ. Continuing, it travels along QSO to T and so on, each time being pulled to the right as it goes up, and to the left as it goes down, and so describing the rosette-shaped path which is dotted in the figure. The actual angle through which the electron is pulled at each vibration is very much exaggerated in the figure. Even with a very strong magnetic field there are many thousands of loops in the rosette, so that the loops are extremely close together and are practically straight lines. We see, therefore, that an electron which started vibrating in the

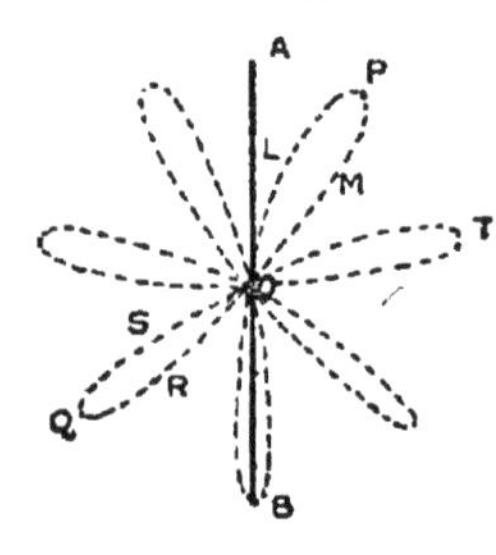

FIG. 48.

ine BA would vibrate in practically a straight line the whole time, but the line itself would gradually turn round. An electron at the point where the light enters the transparent substance would therefore start vibrating along BA; by the time it handed on its vibration to an electron in front of it its line of vibration would have turned a little. This other electron would have its direction of vibration turned a little more before handing it on to the mext, and so on to the end of the substance. The plane of vibration, on emerging, will therefore be turned through an angle which will depend upon the strength of the field and the thickness of the substance.

It will be noticed that in this idea of the mechanism by which the rotation is produced we have made no reference to whether the light is travelling upwards or downwards through the plane of the paper. The direction of the force pulling the electron depends simply on the direction of its motion, not on the direction in which its motion is handed on. The plane of polarisation is therefore turned in the same direction, whichever way the light goes through the apparatus in Fig. 47, either right to left or left to right. Lord Rayleigh shows how it is possible to utilise this result so as to arrange an apparatus which will allow light to pass through in one direction, while absolutely refusing to allow it in the opposite direction. The two Nicol prisms in Fig. 47 are placed so that the directions of vibrations which they will transmit are at 45° to each other, say the directions OA and OB in Fig. 49. First pass the light through the Nicol which transmits vibrations OA and then turn on the current in the electromagnet in such a direction that the plane of polarisation is turned in the direction of the arrow. Adjust the current until the light is completely extinguished by the second

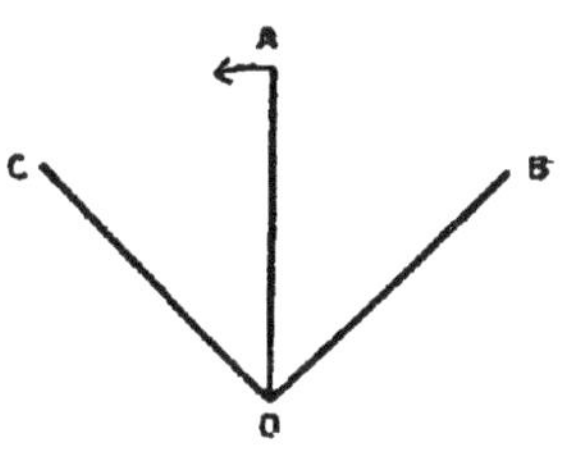

FIG. 49.

Nicol. We know that the vibrations must now be in the direction OC perpendicular to OB, *i.e.* they have been turned through 45° in the direction of the arrow. Now pass the light in the opposite direction. The light will enter the substance with its direction of vibration, OB. On passing through the substance its direction of vibration will be turned through 45° in the direction of the arrow. The direction, on emerging, will therefore be OA, and therefore the light will be completely transmitted by the second prism. We might therefore construct a window through which we might see without being seen.

The Kerr Effect.—Another magnetic effect on light is known as the Kerr effect. Kerr discovered in 1887 that when plane polarised light is reflected at the polished pole of a strong electromagnet the reflected light is no longer plane polarised. It is found that the vibrations have become elliptical, instead of being in straight lines, owing to the reflection at the magnetised surface. The whole theory of the effect is too difficult to be introduced here, but it is mentioned as another piece of evidence to show the intimate connection between magnetism and light.

The Zeeman Effect.—This is perhaps the most important magnetic effect, for it has led to a large number of experiments and to a large increase in our knowledge of the structure of the lines in the spectra of different sources of light. It is the effect of a powerful magnetic field on a monochromatic source of light itself. When such a source is viewed in a spectroscope which produces a large dispersion it is found that a single line has been split up into two or more lines close together. Suppose that in the arrangement for the Faraday effect, Fig. 47, the block of transparent material is replaced by a sodium flame and that a spectroscope is pointed at the flame along the aperture in one of the poles. In this position only those vibrations in the flame which are perpendicular to the magnetic field are effective, for we shall be looking along the magnetic lines of force. Any electron in the flame which would vibrate backwards and forwards perpendicular to the field before the field is established will describe the rosette-shaped path shown in Fig. 48 when the field is established, and for exactly the same reason. Now a simple to and fro vibration is exactly equivalent to two equal circular motions of the same period, but in opposite directions. This can be seen very readily, for if two circular vibrations be impressed at the same time on a particle the actual displacement is the sum of the two individual displacements. Suppose that the two circular vibrations are exactly together at A, Fig. 50, and travel round the circle in opposite directions at exactly the same speed. At A the displacement produced will be twice OA, for each would produce a displacement OA. When one vibration has reached B, the other will have reached C, and each will produce an upward displacement OD. The sideways displacements will however be equal and opposite (BD and CD), and therefore the resulting displacement is equal to twice OD. When one displacement has reached E, the other will have reached F, and evidently the resulting displacement is zero. Continuing in the same way through a complete vibration, it is evident that a simple up and down oscillation results.

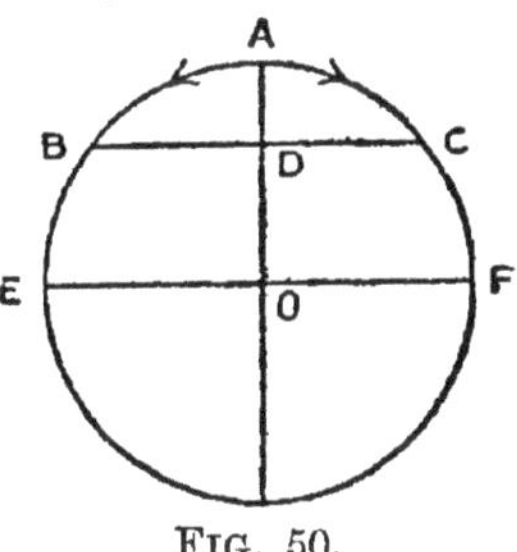

FIG. 50.

Suppose now that the speed of one of the circular components is reduced a little and the other one increased, and suppose they are exactly together at A, Fig. 51. Then, since the two are very nearly similar, they will produce very nearly an up and down oscillation, but not quite. Suppose the vibration going to the right from A is the faster. Then the two will not reach B together, but will be together at some point P to the left of B, the faster vibration having gained a little on the slower. In the same way, getting back to A, the two will be together at some point Q to the right of A, the faster one having gained a little more. At each successive vibration the point of coincidence will be a little more to the right of A and a little more to the left of B, and we shall get a rosette-shaped path exactly like that in Fig. 48. The magnetic field has therefore converted what was equivalent to a simple to and fro vibration, or to two exactly equal circular vibrations, into what is equivalent to two circular vibrations of slightly different periods, one a little faster and the other a little slower than the original period.

The spectroscope will separate these two different periods and show them as a double line, one line each side of the original position. If the spectroscope does this we shall conclude that the vibrating electrons which are sending out the light are executing simple to and fro motions in a straight line, though the direction of the line is different for different electrons.

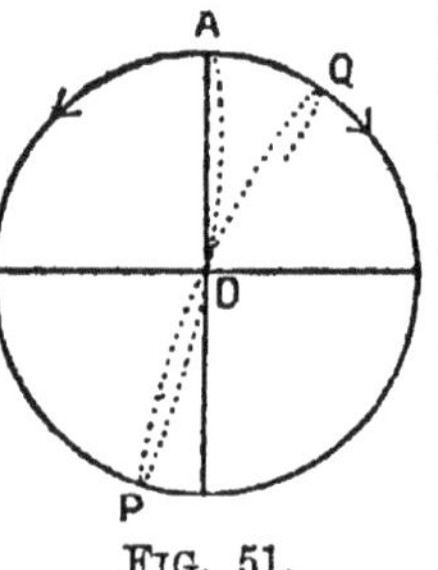

FIG. 51.

If instead of looking through the pierced pole we point the spectroscope at the flame from a point at the side perpendicular to the magnetic lines of force, we get another possibility to the vibrations. Beside the vibrations perpendicular to the magnetic field, we have those parallel to it, for these are now perpendicular to the direction in which the light is travelling. The two circular vibrations of slightly different periods will be produced as before, but we shall now be looking edge on to them and so they will appear to be simple to and fro vibrations and therefore plane polarised. The vibrations along the field are quite unaltered by the field. The original vibration will therefore be split up into three plane polarised components: one will have the same frequency as the original, one will have a little smaller frequency, and the other a little greater frequency. We shall therefore get the one line split up into three in the spectroscope, the middle line being undisplaced.

This case of a simple to and fro motion of the vibrating electrons producing a doublet when viewed along the magnetic field and a triplet when viewed perpendicular to the field is the simplest Zeeman effect. If the vibrations are more complicated the magnetic field divides the line into a larger number of parts, but the statement of the simplest case is sufficient to give an idea of the relationship between the light and the magnetic field.

The Velocity of an Electromagnetic Wave.—Of all the evidences that light consists of short electromagnetic waves, perhaps the most conclusive is the calculation by Maxwell of the speed of an electromagnetic wave. From the well-investigated laws governing electric and magnetic forces and their interaction Maxwell calculated that the speed of electromagnetic waves in air was about 300 million metres per second. Now the speed of light has been very carefully measured by many experimenters, and is almost exactly equal to this quantity. Such a coincidence would hardly be credible unless the two things were identical.

CHAPTER VIII

THE NATURE OF WHITE LIGHT

We have now come to the conclusion that light consists of short, transverse, electromagnetic waves, and we have tacitly assumed a certain amount of regularity in the waves. When a grating or a prism produced light which was indistinguishable from sodium light we have assumed that the period of vibration in it is the same as that for sodium light, and experiments show that it interferes in exactly the same way. In the sodium flame we must conclude that the vibrating electrons which are causing the emission of the lignt vibrate to and fro quite regularly for a number of complete vibrations and so send out a train of exactly similar waves before any disturbance such as a collision between the molecules causes them to break off suddenly and start sending out another series of waves. In a gas the molecules are so far apart that there is time for quite a large number of undisturbed vibrations between successive collisions, and so we might expect long trains of similar waves from incandescent gases. The question naturally arises, Are these regular trains of waves present in the white light, or have they been manufactured by the prism or grating? Newton's experiments at first sight would lead us to the conclusion that all the different regular trains of waves represented in the spectrum existed already in the white light, and that the sole function of the prism or grating is to separate them; but in reality Newton's experiments give no evidence either one way or the other, for the original disturbance might be an irregular one or a mixture of a number of regular ones, and the resulting wave might still be the same when it reached the grating or prism.

Experimental evidence has something to say on this subject, however. Fizeau and Foucault obtained distinct interference fringes with a narrow line which they isolated from white light when the difference of path was 50,000 wave-lengths. By most people at the time this was held to prove that this coloured component in white light vibrated regularly without any sudden change for at least 50,000 complete periods, but it has been found since that the limit to the number of wave-lengths path difference is determined simply by the resolving power of the spectroscope used to detect them, every increase in the resolving power being accompanied by a corresponding increase in the possible path difference. The fact is that interference fringes can be explained just as well on the assumption that they come from a single vibration or pulse, as from a whole train of regular vibrations. Suppose, for instance, that a single pulse is reflected at Fresnel's mirrors and is then received by the prism of a spectroscope. If there is a path difference the pulse reflected from one mirror will reach the prism before that from the other. Each pulse as it enters the prism will start the electrons in it vibrating, and these vibrating electrons of all periods will transmit waves, and so the single pulse gets drawn out into trains of waves of all periods. The second pulse arriving after the first will be drawn out in the same way as the first, but it must arrive an odd number of half-periods later than the first for some of the waves, and so the vibrations of the electrons of these periods would be stopped, *i.e.* we should get interference.

If we suppose that the sensation of sight is due somehow to the vibrations of electrons in the retina, the retina itself will do instead of the prism for drawing out a pulse into waves, and so we may have interference even without the prism. We see therefore that it is just as simple to imagine that the regular trains of

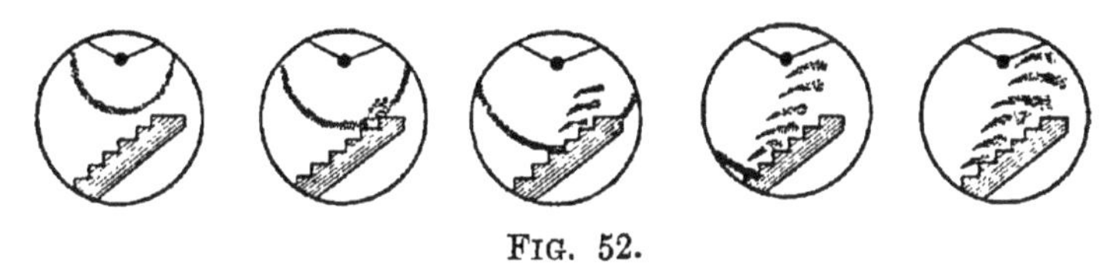

FIG. 52.

waves are produced by the receiver as by the transmitter of the wave. We only need assume regularity of period in one or other of them.

R. W. Wood has given a very pretty illustration of the way in which a grating produces a regular series of waves from a single pulse. He produced a single pulse of sound by means of an electric spark, just as he did in his illustrations of Huygens' wavelets and of reflection. In the path of the pulse he placed a series of little steps so that the pulse was reflected from each step successively. By the same method as before he photographed the pulse in successive stages in its development. The five diagrams in Fig. 52 are drawn from Wood's photographs. The small black circle at the top of each diagram is the shadow of the knobs between which the spark passes which produces the sound pulse. In the first diagram the sound pulse is seen approaching the flight of steps. In the second a little of the pulse has been reflected from the top step.

In the third diagram portions of the pulse have been reflected by three steps, and the three reflected portions are seen following one another out. Diagrams 4 and 5 show more of the reflected pulses following as the original pulse impinges on more of the steps in succession.

By the reflection of a single pulse at a series of steps we have thus converted it into a series of pulses following one another at regular intervals, *i.e.* into a regular train of waves.

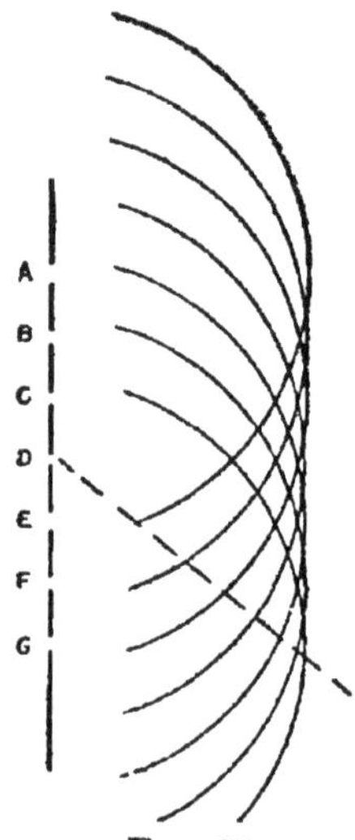

FIG. 53.

The grating will act in the same kind of way. Suppose ABCDE, Fig. 53, are some of the slits in a grating, and suppose a single plane pulse has come up to the grating so that all parts of the pulse reach the grating at the same time. (This is not a necessary assumption, but makes the diagram simpler.) Then a small portion of the pulse is transmitted by each slit and spreads out as a Huygens' wavelet on the further side. Drawing these wavelets at any instant we see that in any direction such as the direction of the dotted line the pulses from the different slits are following one another at regular intervals, the intervals being longer the greater the obliquity of the direction. In the different directions we shall therefore obtain the different colours in exactly the same way as has already been worked out for the grating. It is evident, however, that this is not a case of interference. We shall get light in all directions, and the period will alter with the direction, the intensity gradually falling off as the obliquity becomes greater. If a receiver which absorbs all the waves equally, *i.e.* which has no special period of its own, is placed to receive the pulses it should show no sign of interference fringes whatever. Such a receiver is Langley's blackened bolometer strip. When the region in which interference bands might occur in the white light is explored with a bolometer strip there is very little sign of maxima and minima, but there is a little.

There is a central maximum which has a minimum on each side, then rising to a maximum, and gradually falling off from the second maximum. The intensity curve is something like that in Fig. 54. The fact that there is any sign of interference at all shows that there is a certain amount of regularity in the disturbance, *i.e.* a certain preponderance of certain wave-lengths over others, but the regularity is evidently very slight.

Another fact which would lead us to conclude that there is a certain amount of regularity in the original vibrations which produce white light is that a full radiator shows a perfectly definite distribution of energy in its spectrum, which depends in a perfectly definite way upon the temperature. The full radiator produces white light when raised to a sufficient temperature.

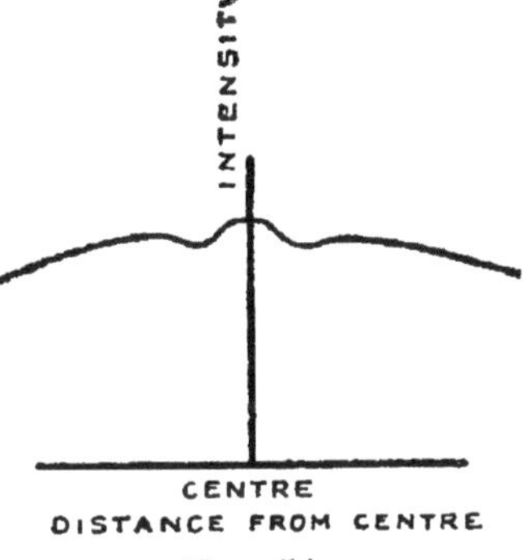

FIG. 54.

A definite distribution of energy might be explained by assuming pulses whose displacement curves have a certain shape and assuming that the shape depends upon the temperature, but it is perhaps simpler to suppose that instead of the simple pulses we have very short trains of waves given out by electrons of all sorts of periods.

When a gas becomes incandescent the molecules are so far apart and their collisions so infrequent compared with the frequency of vibration of the electrons that between two consecutive collisions of a molecule the electrons may execute a large number of perfectly free vibrations. But when a solid or liquid is incandescent (and it is incandescent solids and liquids which produce white light) the collisions between adjacent molecules are so frequent owing to their being so close together that an electron can complete at the most a very few complete vibrations between successive collisions. Beside this we may suppose that the molecules are so close together that the electrons in adjacent ones appreciably affect one another and alter their period of vibration. Thus we may get all variations of periods even though only a few types of electrons are concerned in the emission of the light.

In conclusion, there are three ways in which white light might be produced: (1) by electrons of all periods executing long series of vibrations between successive collisions or any other cause of sudden change of phase; (2) by electrons giving out only single pulses, but of definite shape; (3) by electrons giving out a few vibrations each of all periods.

Any of these three would account for the observed results. The objection to the first hypothesis is that if such a long series of vibrations were executed it is clear that neighbouring molecules cannot interfere appreciably with one another, and consequently we have to assume an immense number of different types of electron in order to account for the waves of all lengths. We should require at least 30,000 different ones to account for the different waves in the visible spectrum alone, for if there were less we should be able to detect gaps between the different waves in the spectrum. The number of types required to explain the whole spectrum would therefore be enormous.

The objection to the second hypothesis is that it is not at all certain that a single pulse can be regularly refracted.

It is therefore better to suppose that the third hypothesis is the true one, and that there is a little regularity in the vibrations of the source, but nothing like the amount which the first hypothesis would suggest.

INDEX

Printed by BALLANTYNE, HANSON & CO.
Edinburgh & London

3/13

15

Milton Keynes UK
Ingram Content Group UK Ltd.
UKHW022031190124
436367UK00004B/223